PR/
THE ENTREF

MW01595880

Jim Hunt's 27 years in local elected office is a testament to his visionary leadership and amazing entrepreneurial spirit. He started early as a young adult by embracing a new technology to launch his own business and continues to see possibilities where others see risks. His ability to transform challenges into opportunities has left a lasting impact on our cities throughout the nation. A Past President of the National League of Cities, Jim's legacy is one of innovation, resilience, and unwavering dedication to public service. Do you want your city to be Amazing? Do you wish your city was more entrepreneurial? Good! This book is for you!

Mike Conduff, Author of the "On Target Board Member"
and President & CEO of The Elim Group

"The Entrepreneurial City is a groundbreaking guide for city managers and elected leaders who are ready to embrace innovation in civic management. With a modern, tech-forward approach, this book masterfully explores how cities can harness social media and digital tools to drive engagement, transparency, and smart growth-oriented strategies for residents and stakeholders alike. Packed with actionable insights, it's an essential read for anyone looking to elevate their city's potential through smart, digital-first strategies."

Sam Toles, Founder CiviSocial, former Chief Content Officer at Bleacher
Report, Former GM of Vimeo, Former Head of Digital at MGM Studios

As a former city councilman and now an advocate for equity, environmental justice, electric vehicles, and sustainable infrastructure with an emphasis on correcting the injustices experienced in underserved communities, I recognize the critical need for forward-thinking leadership in our cities. Jim Hunt's "The Entrepreneurial City" is a blueprint for how local governments can embrace innovation and drive meaningful change. This book is a must-read for any city leader committed to building a sustainable and prosperous future."

Rap Hankins, President of Drive Electric Dayton
and Former Trotwood, Ohio City Councilman

"Jim Hunt provides an excellent blueprint for cities to foster innovation through collaborative entrepreneurial strategies, ensuring they become more resilient, sustainable, and adaptable in a rapidly changing landscape. A must-read for anyone in the private or public sector committed to driving their city's future success."

Nicole Rongo, Vice President of Government Relations
& Strategic Partnerships for CGI Digital

"Innovation and entrepreneurial leadership are less about technology, but more about your mindset and the mindset of your organization. "The Entrepreneurial City" guides us to a path that we should all take to create communities that will be ready for the next set of challenges. Jim Hunt is a must-read for all local government professionals, regardless of position and tenure."

Lee Feldman, Senior Advisor at Zencity and
Past President of ICMA

"I've had the pleasure of knowing Jim Hunt for over twenty years, dating back to our time working together at the National League of Cities. Jim's leadership and vision have always been ahead of the curve, and his approach to city government is innovative and deeply rooted in his belief that local leaders must think entrepreneurially to solve today's complex challenges. No one knows cities better than Jim. His book, *The Entrepreneurial City*, is a must-read for anyone committed to driving change and improving the quality of life in their communities. Jim's ability to draw from his decades of experience in local government paired with forward-thinking strategies make this a powerful guide for city leaders everywhere."

Cathy Spain, Economist and
Senior Director Bearing Advisors

"Branding is not just about logos and slogans; it's a strategic asset that drives economic development, builds trust, and fosters community pride. In an entrepreneurial city, a well-crafted brand can attract talent, inspire civic engagement, and create a sense of belonging that propels the community and economy forward. As I have personally seen in our client's cities like Waxhaw, NC and Aledo, Illinois, a strong brand can transform challenges into opportunities, making branding an essential tool for any city looking to thrive in today's competitive landscape."

Jeni Bukolt, Founder & CEO of HAVEN Creative

"Jim Hunt's insights into the future of city governance are a must-read for local leaders navigating today's complex challenges. "The Entrepreneurial City" provides practical, innovative strategies for addressing issues such as homelessness, infrastructure, and technology integration. The book is an urgent call to action for all who care about the future of their communities."

Senator Elvi Gray Jackson, Alaska State Senate and
Former member of the Anchorage Assembly

"This book encourages leaders to embrace transparency and collaboration. there is nothing more unifying and empowering than to allow citizens and staff access to information and a thorough understanding of the challenges their cities face. I appreciate how Jim communicates the imperative to involve all stakeholders in addressing seemingly intractable problems. we are all in this together!"

Janice Jackson, former City Manager for Augusta, GA
and Host of the Local Matters Podcast

THE
ENTREPRENEURIAL CITY

Marla,
All the Best!

Jim Hird.

THE
ENTREPRENEURIAL CITY

BUILDING SMARTER GOVERNMENTS
THROUGH ENTREPRENEURIAL THINKING

JAMES C HUNT

The Entrepreneurial City
Copyright © 2025 by James C. Hunt
Published by: Amazing Cities Publishing
ISBN: 979-8-9914426-0-2

This book is dedicated to.
Ali, Avery, Emery, and Sadie

*I hope that your lives are
Amazing in everything you do!*

Contents

FOREWARD

By Mayor Steve Patterson of Athens, Ohio

In an age of rapid urbanization, where over half of the world's population now resides in cities, the role of municipal governments has never been more crucial. Our urban centers are the beating hearts of modern civilization, where economies are built, cultures converge, and innovations emerge. Yet, as cities continue to grow and evolve, the traditional models of governance struggle to keep pace with the dynamic challenges of the 21st century. From economic inequality and infrastructure decay to climate change and digital transformation, the complexities facing today's cities are as diverse as they are daunting. It is in this context that Jim Hunt's latest book, *The Entrepreneurial City: Building Smarter Governments through Entrepreneurial Thinking*, provides a timely and insightful blueprint for reimagining municipal governance.

Jim Hunt is no stranger to the intricate interplay between entrepreneurship and governance. As a seasoned elected local leader, state housing official, community revitalization expert and business consultant, Hunt has spent decades at the forefront of city development, bringing innovation and technology to local government organizations. His unique perspective, cultivated through years of hands-on experience, forms the foundation of this book. With *The Entrepreneurial City*, Hunt invites us to explore the possibilities that emerge when governments adopt the mindset, strategies, and agility of entrepreneurs.

In this book, Hunt lays out a compelling argument for why the future of cities depends on the ability of their leaders to think and act like entrepreneurs. But what does that really mean? In essence, it's about embracing risk, fostering innovation, and cultivating a spirit of experimentation—traits traditionally associated with the private sector.

Hunt argues that these traits are not only applicable but essential to the public sector, particularly at the municipal level. When cities are managed with the same creativity and resourcefulness as successful startups, they can become more resilient, adaptable, and responsive to the needs of their citizens.

Hunt's vision of the entrepreneurial city is not a call for cities to abandon their public responsibilities in favor of profit-driven motives. Rather, it is an invitation to rethink how municipalities can better serve their residents by leveraging entrepreneurial principles. At the core of this vision is the belief that cities should be driven by a commitment to solving problems in innovative ways, rather than being bogged down by bureaucratic inertia. Hunt illustrates how entrepreneurial thinking can lead to smarter, more efficient, and more equitable urban governance by offering practical examples from cities across America that have successfully implemented these ideas.

One of the most compelling aspects of *The Entrepreneurial City* is its focus on real-world applications. Hunt goes beyond theory to examine case studies of cities that have embraced entrepreneurial approaches to governance, resulting in significant improvements in areas such as public transportation, housing, and sustainability. From the tech-driven transformation of Union City, Georgia to the community-focused initiatives in Tacoma, Washington, Hunt demonstrates that entrepreneurial thinking is not a one-size-fits-all solution, but a versatile toolkit that can be adapted to meet the unique challenges of any city.

Another strength of this book is Jim's emphasis on the importance of collaboration. He underscores that the entrepreneurial city cannot be built by governments alone. Instead, it requires a multi-stakeholder approach that includes businesses, non-profits, academia, and, most importantly, the citizens themselves. By fostering partnerships across these sectors, cities can pool resources, share expertise, and co-create solutions that are both innovative and inclusive.

As we navigate the complexities of the modern urban landscape, *The Entrepreneurial City* serves as both a guide and a call to action for city leaders, policymakers, and citizens alike. James Hunt challenges us to rethink the role of government in our cities and to embrace the entrepreneurial spirit as a force for positive change. By doing so, he offers a path forward for cities to become not only smarter and more efficient but also more humane and just.

In reading this book, I am reminded of the words of the famed urbanist Jane Jacobs, who once said, "Cities have the capability of providing something for everybody, only because, and only when, they are created by everybody." Hunt's work is a testament to this idea, showing us that by adopting an entrepreneurial approach, we can create cities that are vibrant, resilient, and truly inclusive. It is with great pleasure that I introduce *The Entrepreneurial City* and commend Jim for his invaluable contribution to the ongoing conversation about the future of urban governance.

This book is not just a vision of what our cities, towns and villages could be; it is a roadmap for getting there. As you turn these pages, I encourage you to reflect on how the principles of entrepreneurship can transform not only our cities but the very way we think about governance itself.

PREFACE

The inspiration for "The Entrepreneurial City: Building Smarter Governments through Entrepreneurial Thinking" began to take shape in 2018, while I was writing a column for the Harrison County Journal. At the time, I was consulting with several companies serving local governments, and I couldn't help but notice the contrast between the private sector's eagerness for innovation and the hesitance of many government officials to embrace new ideas.

In many cases, local officials seemed hesitant to implement new ideas, viewing them as too risky—even when these innovations had the potential to save significant tax dollars. Maintaining the status quo felt like "zero risk," while experimenting with new approaches appeared to threaten their positions or even their chances for re-election.

As someone who served as an elected official for over 27 years, I empathize with their reluctance. I understand the urge to stick with familiar methods. Yet, I also believe that we cannot afford to let fear of the unknown prevent us from exploring solutions that could dramatically improve our communities.

This book is the culmination of my reflections on these challenges and my conviction that local governments can thrive by adopting an entrepreneurial mindset—one that embraces innovation, calculated risks, and strategic partnerships to bring about real change. By adopting innovative practices and forming strategic partnerships, local governments can not only save money but also significantly improve the quality of life for their citizens.

In "The Entrepreneurial City", I explore various challenges facing local governments today, such as homelessness, drugs, crime, zoning disputes, and public safety. I delve into how these problems can be approached with entrepreneurial thinking, leveraging technology and innovation to create more effective and efficient solutions.

Furthermore, this book discusses the role of businesses in assisting cities through partnerships and agreements outside traditional city hall solutions. It suggests practical strategies such as using brand licenses, cooperative agreements, appointing innovation officers, and creating business incubators to engage private sector involvement and attract young talent to city government roles.

You will also read about my experiences with Sunnyside Up-Campus Neighborhoods Revitalization Corporation, where I was able to put entrepreneurial thinking to work in a tangible way. Sunnyside Up became a living laboratory that not only energized a neighborhood but also sparked hundreds of millions of dollars in investment. Through this experience, I learned the immense value of identifying 'champions' and business partners to drive meaningful change.

I hope this book serves as a practical guide for local government leaders ready to embrace change and seek innovative solutions to the pressing issues of our time. Through entrepreneurial thinking, we can build smarter, more resilient governments that are not just equipped to tackle today's challenges, but also to thrive in the future.

ACKNOWLEDGEMENTS

First and foremost, thank you to my wife, Pam—my best friend, rock, and constant sounding board throughout the process of putting this book together. You have filled in the gaps in our busy lives, providing strength and balance when I needed it most. I couldn't have done this without you.

To my daughter, Sarah, and her husband, Matt, and to my son, Jason, and his wife, Stacie—your love and unwavering support have been a constant source of inspiration. And to my four granddaughters, Ali, Avery, Emery, and Sadie—you are the light of my life, reminding me daily of the importance of building a brighter future for the next generation.

To my colleagues at Bearing Advisors—Phil Riley, Brad Carmichael, Mike Conduff, Cathy Spain, and Mike Madden—it has been an incredible journey working alongside you. I've learned so much from each of you, and I am grateful for the wisdom, insights, and shared experiences we've built together.

A special thank you to the National League of Cities, particularly Executive Director Clarence Anthony and your extraordinary staff. You have become like a second family to me over the course of my 39-year relationship with this remarkable organization. The friendships I've made are among the greatest treasures of my life, and it is an honor to release this book as NLC celebrates its 100th anniversary.

To the West Virginia Municipal League, you've been my touchstone, offering inspiration and encouragement when I needed it most. The relationships I've formed here—both past and present—are ones I will always cherish.

Finally, to the City of Clarksburg, my lifelong home, you've shaped me, and I hope in return I've made you proud.

INTRODUCTION

On July 1, 1985, I stepped through the arched brick doorway of the Clarksburg Municipal Building to be sworn in as a council member. At that moment, as the youngest member of the City Council of Clarksburg, West Virginia, I was unaware that this ceremony would mark the beginning of a lifelong journey through the corridors of local government—a journey that would eventually take me to lead the National League of Cities as President in 2006 and visiting over a thousand city halls across the United States and beyond.

Growing up in Clarksburg, my connection to city government was deep-seated. My father, Rex Hunt, was a heavy equipment operator for the Clarksburg Public Works Department and the union president for the city's public works employees. Every night, he shared stories of his day, weaving politics seamlessly into conversations about his job. Additionally, my acquaintance with Jim Blair, a city councilman in Clarksburg and the son of my employer at Blair Motor Supply, where I worked during college breaks, further piqued my interest in local politics. I followed Jim's actions as a city councilman in the local paper and enjoyed his visits to the store to discuss local politics. Impeccably dressed, he epitomized my vision of what a local leader should be.

After college, I was offered a position as a Learning Disability Specialist for our local school system. I loved the opportunity to work with kids and help them along in their studies. I probably would have stayed in education until retirement if it had not been for a fateful decision to become an entrepreneur at the age of 26. I purchased a franchise for an auto rustproofing service, put up everything I owned as security for a Small Business Administration loan, and began a journey of entrepreneurship that continues to this day.

While working in my shop one day, a fellow stopped in and asked if I had ever thought about getting involved in local government. I said that I had given it some thought but was too busy with my business to pursue it. I went home that night, and it wouldn't leave my mind, and I decided to throw my hat in the ring for a seat on city council. Running for office as a young person was a challenge since few people knew me, but after a couple of tries, I was elected as a city councilman in 1985.

I was excited to begin my career as a city councilman. A few days before my first meeting, a large manila envelope was delivered to my house by a city policeman. It was the agenda for the meeting, along with various reports and support materials. After spending a few hours reading the materials, I was struck by the difference between my small business and the operation of a major West Virginia city. While personal computers were in their infancy, I had a Tandy 1000 computer in my office and used it for many timesaving tasks, like letters, bookkeeping, and graphic design. When I went to the city, I saw that they used typewriters for letters and reports without a computer in sight. They even used the travel agency, located across the street, to send and receive faxes. I almost expected to see the Pony Express ride up at any minute.

As I got my bearings as a city council member, I began to look for ways to improve our city. Operating my own business, I realized that the city operated vastly differently than the private sector. Although we served the public, there was little desire to do things differently and a lot of pushbacks from employees and the public if we tried to innovate. While computers were becoming ubiquitous in private businesses, local government still operated largely with pencil and paper.

During my 27-year career as Mayor and City Councilmember in Clarksburg, I watched the growth of technology in business and industry, but local governments seemed to lag in efficiency and innovation. The city staff had not had the opportunity to learn about changes occurring and little effort was made to send staff members to training workshops and conferences where they could keep up with rapid changes in technology. While businesses were experimenting with computing and the growing digitization of tools and machinery, cities were stuck in the past for a myriad of reasons.

In today's world of smartphones, supercomputers, and artificial intelligence, some might find it hard to imagine operating a city without these tools. Rarely did cities and towns venture beyond their narrow silos, and working with private entities was often scorned by both the governing bodies and the public. Public-private partnerships (PPPs) did not begin showing up till around the end of the 20th Century, and even then, they were often looked at with skepticism.

Entrepreneurial cities recognize that it is impossible to operate without establishing mutually beneficial partnerships. Indeed, few cities have the resources and expertise to keep up with an increasingly complex world. Services like data analysis have become as useful for local governments as they are for Walmart and Amazon. Sales and property tax data can predict trends and opportunities for cities and towns and aid economic development strategies. Police departments employ a host of technologies that dwarf the handcuffs and Billy clubs of years past. Modern law enforcement has access to drones that can serve as first responders, assessing tactical responses and protecting officers' lives. Fire departments can access detailed floorplans and building contents as soon as they receive a call, dispatching the appropriate manpower and equipment, saving precious time and protecting lives. Water and sewer utilities use underground cameras and detection devices to find possible weak spots in pipes and areas of erosion and movement. City managers and other administrative staff can prepare accurate strategic plans and projections for all aspects of city operations.

The question that constantly frustrated me as a public official was why cities failed in some of their most basic jobs. As businesses left our struggling downtown, we seemed more willing to harass property owners than look for innovative solutions to fill these buildings. Cities, it seemed, were stuck in a rut. We would hear from the business community that cities should run like a business, but cities weren't simply businesses churning out profits, and their stockholders—the citizens—resented spending tax dollars on a vacant downtown when sidewalks and parks were their immediate need.

What cities craved was the audacity, the risk-taking spirit that seemed plentiful in the private sector. Maybe, just maybe, we could inspire this type of spirit within the walls of city hall. I met such a person from a struggling coal town in the hills of Appalachia.

Amber Miller-Belcher is the city clerk of Logan, West Virginia, and her Mayor, Serafino Nolletti, recognized that his city clerk in stiletto heels was a force to be reckoned with. A successful businesswoman in her own right, Amber injected the entrepreneurial spirit into the city, from city hall to the folks who hung out on the wall outside the courthouse. Her positive attitude was infectious, and when problems arose, she was at her best.

One Friday afternoon, the city of Logan was filled with the sounds of sirens as the Fire Department received a call that an abandoned building had collapsed. When they arrived, a three-story brick building was teetering, with over half of it lying in piles of rubble in the street. As they assessed the situation, it seemed a final nail was driven into the heart of the struggling city. Heads shook as the local gadflies prepared the obituary for the city. Not Amber. She immediately announced that the city had found its newest greenspace—a place with trees, flowers, and benches in one of the best spots in town. All they had to do was remove tons of bricks and finish demolishing the old structure. As many laughed, Amber organized the city to partner with anyone who would listen to her vision. Within months, the debris was gone, green grass covered the site, and the city had a new downtown park. They recently painted murals on the brick walls, and it is a hopeful light in this gutsy city. This was the entrepreneurial spirit that I was looking for—someone who, when faced with seemingly insurmountable challenges, hustled, pivoted, and found solutions that others had missed.

Why are there not more of these inspiring people working in local government? The simple answer is that these firebrands are often seen as troublemakers, their innovative ideas extinguished by bureaucracy. The Entrepreneurial City is a roadmap to cultivate these leaders, ignite the entrepreneurial spirit within city walls, and transform our communities.

In the following chapters, I will introduce you to more of the inspiring people leading the entrepreneurial movement in America's cities and towns. We will also learn the techniques and strategies needed to adopt this mindset in your city or town. Some concepts in the book will challenge your thinking and bring clarity to this new philosophy. It will not be easy to abandon the status quo, but I hope you keep an open mind and learn from hundreds of trailblazers.

CHAPTER 1

WHAT IS AN ENTREPRENEURIAL CITY?

> *"We cannot solve our problems with the same thinking we used when we created them."*
> **Albert Einstein**

An "entrepreneurial city" thrives on innovation in governance, active citizen participation, and a commitment to sustainable growth. It fosters a mindset that encourages staff to look for innovative approaches to solving problems and leveraging technology for the public good. Citizens and businesses are part of the team and are actively encouraged to suggest solutions and identify resources. The approach is proactive and requires buy-in from the governing body, so as not to punish risk-taking, but rather to learn from successes and failures and focus on continuous improvement.

Today's employees look for organizations that value innovation and creativity and cities that actively encourage entrepreneurial thinking can attract talented employees, looking to join a forward-thinking and interesting team. As the private sector deals with labor shortages and employee downsizing, cities can engage employees with real-life challenges and opportunities to create the cities of the future.

To illustrate the power of entrepreneurial thinking in local governance, let's look at the story of Shae Strait. I interviewed Shae, the Director of Planning and Development of the City of Fairmont, West Virginia on my Amazing Cities and Towns Podcast and I was impressed with his excitement for his job. The City of Fairmont is a city in West Virginia that has been slowly losing population over the past fifty years and is home to several "brownfield" sites that housed several glass factories and other industries associated with the coal mining industry. Shae grew up about 20 miles from Fairmont and attended Fairmont State University where he received both his Bachelor of Architecture and his Master of Architecture degrees. He worked as an urban planner in Huntington, West Virginia before coming home to Fairmont. Shae was ecstatic when he talked about his efforts to revitalize the city and put these abandoned factories back on the tax rolls. It would be hard to imagine that a young architecture graduate would have the responsibility and experience that Shae is getting, leading his team in transforming his city from a rusty, industrial town to a sustainable and economically growing community.

The odds are not fully in Shae's favor and some question how this brash young man can undertake such a challenging job when so many have failed. In true entrepreneurial spirit, he converts the doubters one at a time and makes sure that his preparation is flawless. He is also a multitasker and knows that he must keep multiple balls in the air if he is to succeed at his vision. Shae's entrepreneurial spirit shines through in his ability to juggle the revitalization of brownfield sites while simultaneously advancing critical zoning revisions and strategic plans.

As you will see in the coming chapters, creating an Entrepreneurial City has many moving pieces. Citizens need to be engaged as participants and collaborators. Gone are the days of talking "at" people, to a new model of talking "with" people. Citizens in Entrepreneurial Cities can utilize online platforms and town halls to suggest better ways of serving the public. They will be encouraged to be part of the solution instead of just identifying the problems.

In conclusion, an entrepreneurial city is built on three core elements: innovation in governance, active citizen participation, and sustainable growth. It creates a dynamic and innovative ecosystem that actively engages its citizens in shaping a vibrant urban future. It leverages technology, fosters public-private partnerships, and maintains an open dialogue with its residents to ensure that the collective intelligence and creativity of the community are harnessed. Such a city prioritizes sustainable growth, inclusivity, and resilience, adapting swiftly to new challenges and opportunities. By cultivating a spirit of entrepreneurship within its governance and encouraging active citizen participation, an entrepreneurial city not only enhances its economic prospects but also improves the quality of life for all its inhabitants. This approach transforms urban centers into hubs of innovation and cooperation, setting a robust foundation for continuous improvement and prosperity.

Is your city ready to become an entrepreneurial city? What steps will you take to make this transformation a reality?

CHAPTER 2
DEFINING ENTREPRENEURIAL LEADERSHIP

"Innovation distinguishes between a leader and a follower."
Steve Jobs

E ntrepreneurial leadership, within the context of an entrepreneurial city, involves applying entrepreneurial principles and practices at the municipal or community level to foster innovation, economic growth, and social vitality. Such leadership is characterized by proactive, innovative, and risk-tolerant approaches to urban management and development, aiming to transform cities into hubs of opportunity and innovation. In an entrepreneurial city, leaders must embody these principles to drive economic vitality and social innovation.

Over my career in local government, I have worked with many entrepreneurial leaders. Past National League of Cities President, Mayor Tony Williams was someone who epitomized a visionary leader who always seemed to be looking ten steps ahead. I recall a bus tour in Washington where Tony served as our tour director, taking a group of National League of Cities officials throughout the city. I marveled at his knowledge of the city and the areas of opportunity that he seemed to be placing on his city chessboard, just waiting for the stars to align. His work to locate the Washington Nationals baseball franchise along the Anacostia River in the Navy Yard neighborhood of Washington, DC is one of his biggest achievements. I was with him on several occasions when he would be on the phone in heated discussions with developers and financiers to hammer out details of the nearly $1Billion dollar ballpark. Mayor Williams' ability to envision a revitalized Anacostia Riverfront, anchored by a billion-dollar ballpark, exemplifies entrepreneurial leadership—turning a bold vision into a transformative reality for the city.

Mayor Williams was also a visionary when it came to his tenure as President of the National League of Cities. It was customary for each President to pick a topic to be explored when they were the President-Elect and then during their presidential year, the subject would be introduced to the membership, and implementation of the topic would begin. Tony had the idea that the National League of Cities should develop the tools to become the "Google for Cities"! The year was 2004 and Google was just beginning to be a force in the online search business. It was a huge task, and the National League of Cities expended resources and manpower to research and prepare a plan to implement Tony's vision. The vision was bold, but it intersected with Google's growth in the marketplace, where they had just introduced Gmail in 2004 and Google Maps in 2005.

In line with the entrepreneurial spirit, although the goal was not attained, the National League of Cities gained valuable knowledge and perspective that serves it even now. Since 2012, Tony Williams has served as chief executive officer/executive director of the Federal City Council. His tenure as Washington, DC mayor has been hailed by local government leaders, national commentators, and historians, as one of the most consequential mayors in U.S. history.

Another entrepreneurial leader that I came to know during my time as an officer for the National League of Cities was Mayor Rick Baker of St. Petersburg, Florida. Rick wrote *"The Seamless City-a Conservative Mayor's Approach to Urban Revitalization That Can Work Anywhere"* which was published in 2011.

I was in St. Petersburg for a NLC meeting and Mayor Baker took us on a tour of the city. St. Petersburg is a historic and beautiful city, but the mayor decided to focus on seeing the apartment complexes that were built under the federal HOPE IV program. I was sitting in the front of the bus with the mayor, and we arrived at a senior living center and filed off the bus for a quick tour. As we finished the tour we were getting back on the bus, I saw Rick get on his cellphone. He had called his housing authority director to report a broken screen on one of the units. I said, "Being the mayor means you're never off the clock!" and he replied quickly, "If we let things like that go, it spreads!". Mayor Baker's quick action on a minor issue reflects his broader entrepreneurial mindset—one where small detail is as critical as large-scale projects in maintaining the city's vitality. Not surprisingly, Rick Baker is now the Director of the University of South Florida's Innovation Partnerships.

Mayor Tony Williams and Mayor Rick Baker are examples of entrepreneurial leadership at its best. Risk takers who protect the taxpayer's investment, as if it were their own Silicon Valley start-up. I think it is important to note that Mayor Williams and Mayor Baker are on opposite ends of the political spectrum, but both recognize the need to do things differently. Regardless of political affiliation, entrepreneurial leaders focus on what works—innovative solutions, risk management, and a relentless commitment to the community's future."

As cities face increasingly complex challenges, the question remains: Will you lead with the entrepreneurial spirit that drives innovation and progress?

CHAPTER 3

CREATING AN ENTREPRENEURIAL ENVIRONMENT

> *"Give your employees the freedom to experiment and make mistakes, and you'll see true innovation."*
> **– Kimball Fisher**

Fostering an entrepreneurial environment in local government is not just about adopting new practices—it's about transforming the very culture of how a city operates, empowering employees, and engaging the community in meaningful ways. It can be uncomfortable and threatening to some members of the team. Acknowledging and addressing these concerns from the outset can help ease the transition and build trust within the team. It should be developed as a group project and one that encourages communication and collaboration, where everyone's voice is heard and valued. It is also a good time to involve members of the public and the business community, so they can be advocates and partners in the venture. It should look and feel different than the status quo and move organically rather than on a strict schedule. This cultural transformation is essential for cities that aim to stay competitive and resilient in an ever-changing world.

I think it is also time to reexamine long-held views about residency requirements and flexible work hours. I used to be very old school about this, but my mind has changed in the past several years when I speak to some of the incredible people who have left local government. Two-income couples are increasingly shut out of local government because of their housing choices, childcare needs, or other issues. I read a recent story about a city manager who had to leave local government when his wife was selected as a city manager in a nearby city. What should have been a joyous time, turned into a nightmare when both cities had requirements that the city manager had to live within city limits. Many cities struggle to fill public safety positions but lose good officers due to the cost of housing in a community or the lack of supply of affordable housing.

Rigid work hours also impact finding and keeping quality employees. Childcare and elder care are often considerations for employees and not moving a starting time or quitting time to accommodate families is often the stated reason for leaving the job. I recognize that this creates some logistical issues but having vacant positions also creates stress in the workplace and makes recruitment all the more difficult.

Traditionally, local government careers have been perceived as lower-paying jobs but with generous pensions and benefits. Unfortunately, many cities have cut these benefits and pensions without offering competitive pay increases. As the Baby Boomer generation retires in record numbers, local governments need to be resourceful and flexible in filling these positions. The "Great Resignation" is here and much of the institutional knowledge that built the city's systems will need to be replaced with a new workforce. I recall a time when our city's Sewer Department was struggling to locate a buried line. After several unsuccessful attempts and a lot of frustration, someone suggested, "Let's call Pat!" Pat had retired years earlier and, despite needing a cane, was always eager to help. Sure enough, when Pat arrived, he hobbled over to an untouched area and confidently said, "Dig here!" After puffing on his pipe, he climbed back into his rusty truck and waved goodbye. One day, Pat is not going to be there to help them!

Here are some ideas to build an entrepreneurial environment in your city:

✦ Gain Buy-In from Elected Leaders

- **Create the Vision**: Identify the benefits of an entrepreneurial mindset, such as increased economic growth, job creation, and improved quality of life. Give examples and data to demonstrate success in other cities.

- **Engage Leaders in the Process**: Explain the process and how they are an integral part of the success. Solicit their input and address any concerns they may have. Be careful not to let leaders "lead" the process but rather be interested participants, on par with everyone on the team.

- **Celebrate Early Wins**: Document your progress and find ways to celebrate small victories. Success breeds success! Ribbon cuttings, groundbreakings, or even a "Donut Day" to celebrate some success can help the team stay focused and realize that their efforts have not gone unnoticed.

✦ Engage All Departments

- **Be Transparent:** Implementing an entrepreneurial environment can be uncomfortable for employees who might be resistant to change or fear for their jobs. Be clear that their input is valued.

- **Encourage Cross-Departmental Collaboration:** Too many times, departments operate in silos without seeking assistance from other departments. When doing projects or training, try to mix up members from various departments. Make it fun!

✦ Engage the Business and Education Community

- **Seek out Business Leaders:** Discuss with the local Chamber of Commerce and get suggestions for business leaders who might make a good resource for the city on specific projects or advisory committees. Speak to local civic clubs or organizations about the effort and encourage their input.

- **Identify Joint Projects:** Collaborate on projects that benefit both the city and local businesses. Clean-ups, community murals, festivals, etc. can create strong bonds and build a network of people to improve the community.

- **Publicize Business/Community Efforts:** Use local papers, social media, radio, and television to highlight how the city and business community are working to create an entrepreneurial environment.

✦ **Encourage Open Dialogue**

- **Establish Communication Opportunities:** Hold regular workshops, trainings, and meetings to allow employees to share ideas and obstacles to building an entrepreneurial environment. Creating a safe space for discussion can lead to breakthroughs in innovation.

- **Celebrate Innovation:** Be generous with recognition for finding innovative solutions to problems faced by the city. Certificates, gift cards, and recognition at council meetings can let employees know that you value their contributions.

✦ **Offer Entrepreneurial Education**

- **Workshops and Training:** Offer workshops and training programs to employees on critical thinking, technology, and innovation. This can grow the entrepreneurial mindset within the workforce.

- **Invite Guest Speakers:** Identify speakers on subjects like innovation, artificial intelligence, or technology to provide regular programs for employee luncheons or training sessions. Public officials from communities that have had success are another great resource and with Zoom availability, distance is not an obstacle.

✦ **Invest in Technology**

- **Tools:** Keep up to date with technology and equip your employees with the tools that they need to improve city operations.

- **Training:** Commit to life-long learning and encourage employees to identify opportunities to hone their skills.

✦ **Encourage Partnerships**

- **With other government entities:** Host other communities in your county or region to visit your city and share best practices.

- **With local businesses:** Visit local businesses to see what they are doing and how they are dealing with issues like cybersecurity, remote work, artificial intelligence guidelines, and other topics.

- **With Colleges and Universities:** Establish relationships with local higher education institutions to see what is on the horizon and how you might gain from collaboration.

- **Think Globally, Act Locally:** Reach beyond your city limits and look for innovative approaches on the state, national, and international scale. Sometimes seeing the problem from other's eyes can reveal a solution.

Like a new pair of shoes, creating an entrepreneurial mindset takes some getting used to. Don't be afraid to fail and use each twist in the road as a learning opportunity. Include everyone and be careful not to alienate members of the team. Each person can contribute and don't assume that just because a person is older, they are not as tech-savvy as younger team members. Avoid jargon. Some teams have made this a fun game and "fined" members for using acronyms and jargon. After a few quarters in the "Jargon Jar", everyone was on the same page and took the time to explain terms used in the tech world.

Creating an entrepreneurial environment in a city is about embracing a culture of innovation and adaptability. It's about transforming every challenge into an opportunity and every idea into a catalyst for growth. As your city evolves, encourage a spirit of collaboration, risk-taking, and creativity among your workers and leaders. By fostering this dynamic ecosystem, you'll not only enhance economic vitality but also inspire a community where innovation thrives, and the future is shaped by those bold enough to dream and act.

AN INSPIRING ENTREPRENEURIAL STORY: I spoke to Ansley Fender, Founder and CEO of Grantcycle on my Amazing Cities and Towns Podcast about her experiences in going from a homeless teenager and classically trained violinist to founding a successful tech startup in Bloomington, Indiana. Ansley embodies the entrepreneurial spirit that can inspire local government employees. Her company, Grantcycle is part of the Mill, a former vacant furniture company that was transformed by the City of Bloomington into a 12-acre technology park. Ansley's success story exemplifies how creating spaces for innovation, like Bloomington's technology park, can empower entrepreneurs and, in turn, benefit the entire community.

Ansley is enthusiastic about how this innovation center helped her launch her business that assists cities, towns, and other governmental organizations in administering grants. Imagine the collaboration that could occur between someone like Ansley Fender and a city wanting to develop an entrepreneurial environment. I give her as an example of the type of person who may be in your community and would love to assist their local government and maybe even learn a thing or two from the city. This kind of collaboration between the public and private sectors is at the heart of an entrepreneurial environment, where both entities can learn from each other and drive mutual success. Cities can actively identify and nurture local talent by creating innovation hubs and providing resources that encourage entrepreneurial growth

CHAPTER 4
FOSTERING ENTREPRENEURIAL ENGAGEMENT

> *"A small group of thoughtful people could change the world. Indeed, it's the only thing that ever has."*
> **Margaret Mead**

Walking into over one thousand city halls in my career, you can almost feel the energy that exudes from an entrepreneurial city. Groups of people working on projects or assisting the public with their needs. No long lines or glassed-in cages where your energy dissipates, and the chatter is about the drudgery of the experience.

I can remember my first of many visits to the City of Bellevue, Washington's City Hall. Artwork on the walls and a coffee cart in the hallway give off the vibe of a tech startup. Comfortable chairs and bright surroundings in the Permit Center, where staff are sitting with businesses and individuals who are discussing their projects with large tables to accommodate plans and documents. The City of Bellevue has effectively blended the "virtual" city hall with the in-person experience and constantly solicits suggestions to make the experience better.

Figure 1 The impressive entrance to the Bellevue, WA City Hall

The design and atmosphere of Bellevue's City Hall not only create a welcoming environment but also fosters a spirit of collaboration and innovation, which are key to entrepreneurial engagement.

Figure 2 The Coffee Shop in the Bellevue City Hall

As you work to instill entrepreneurial thinking in your city, strive to be seamless and inclusive. Remember that everyone starts from a different set of experiences and what works for some may not work for all.

Here are some engagement strategies to encourage citizen involvement:

✦ **Digital Platforms for Engagement:** Develop interactive digital platforms where citizens can submit ideas, vote on proposals, and provide feedback on development projects. There are several platforms available to cities that have positive track records in being accessible and user-friendly. Ensure these platforms are designed with inclusivity in mind, offering features that cater to all age groups and varying levels of technological proficiency.

✦ **Public-Private Partnerships (PPPs):** Bringing businesses to the table can have great value for cities and citizens. Many businesses recognize that their success is tied to the success of the city where they are located. Over my career in local government, there have been very few times when we reached out to the business community when they have not responded. The most common question we were asked is "Why didn't you come to us sooner!" Encourage collaboration through regular meetings with business leaders or creating advisory boards that include key business stakeholders.

✦ **Educational Programs and Workshops:** Citizens are interested in where they live and how the city operates. Providing informative workshops and programs on various aspects of city government is a good way to inform and engage citizens. A tour of the city's water and/or sewer plant can give citizens a chance to understand the increasing complexity of these operations and some of the challenges that the city deals with regularly. A workshop on criminals who prey on senior citizens can inform citizens and let them know about the city's efforts to decrease this type of criminal activity. Empowered with knowledge, citizens can participate more effectively and responsibly. Tailor these programs to address the diverse needs and interests of different demographics within the city, ensuring that everyone feels included and empowered to participate.

✦ **Transparent Communication:** Maintain a transparent approach to governance by openly sharing information about city operations, budget allocations, and the status of ongoing projects. When cities are forthright in their communications, requests through Freedom of Information Requests (FOIA) and outcries at public meetings are reduced. Cities can use online dashboards, newsletters, and social media updates to provide real-time information on budgets, projects, and other key operations.

✦ **Innovation Challenges and Competitions:** Organize city-wide challenges or competitions to solve specific problems or improve local services. These initiatives can harness the creativity and local knowledge of citizens to develop practical solutions. Identify an issue within the city, such as graffiti or high weeds, and solicit innovative solutions from citizens and other groups like college and high school students. Publish the results and invite innovative suggestions to attend a public meeting with staff and discuss why or why not these solutions can work. Create a small pin or coffee mug with the words, "City of Anytown, USA-Innovation Expert" or some other catchy slogan.

✦ **Volunteering:** Create a user-friendly system where citizens can easily find and sign up for volunteering opportunities that help improve the city. This could include environmental initiatives, social programs, and more. Be inclusive and make sure that all citizens are encouraged to participate. Actively reach out to underrepresented groups in the community, ensuring that volunteer opportunities are accessible and welcoming to all.

✦ **Leverage social media, Newspapers, TV, and Radio:** Use all available media to spread the word about volunteer opportunities and celebrate community achievements and events. Make interesting videos that illustrate how the city's infrastructure works. These do not have to be highly produced and are often more effective when they have a local feel with real-life employees. Use storytelling to make your media outreach more engaging. Share personal stories of citizens and employees who have made a difference

Engage, engage, engage! By fostering a culture of engagement, cities not only build stronger communities but also unlock the collective creativity and innovation needed to tackle today's complex challenges.

Take a moment to assess your city's current engagement strategies. What steps can you take today to create a more inclusive, proactive, and entrepreneurial community?"

CHAPTER 5

PROFILES OF ENTREPRENEURIAL LEADERS

"The greatest leaders mobilize others by coalescing people around a shared vision."
– Ken Blanchard

In the realm of local government, entrepreneurial leaders stand out not only for their innovative ideas but also for their ability to turn those ideas into tangible improvements for their communities. As I reflect on my journey through various roles in local governance, I've had the privilege of crossing paths with numerous visionary individuals who embody the spirit of entrepreneurship. This chapter highlights twenty such leaders—mayors, councilmembers, city managers, and city clerks—who have each brought unique perspectives and transformative solutions to their roles. While this is by no means an exhaustive list, it provides a glimpse into how entrepreneurial thinking can reshape the landscape of local government, driving progress and fostering a spirit of dynamic leadership within our cities and towns. These profiles not only celebrate the achievements of these leaders but also serve as a blueprint for how entrepreneurial thinking can be applied in different local government contexts. Through their stories, you'll discover strategies for overcoming challenges, fostering innovation, and engaging communities—all critical components of entrepreneurial leadership. Let me introduce you to these Amazing leaders:

MAYOR STEVE PATTERSON of Athens, Ohio:

At the 2023 National League of Cities "City Summit" in Atlanta, Georgia, I was invited to a dinner with Gayle Manchin, the Federal Co-Chair of the Appalachian Regional Commission, Mayor Steve Williams of Huntington, West Virginia, and Mayor Steve Patterson of Athens, Ohio. I had known Mayor Williams and Mayor Patterson for several years and admired both for their dedication to public service. During the dinner, Mayor Patterson shared a compelling plan with Federal Co-Chair Manchin. His vision was to connect mayors from across Appalachia to share best practices and learn from one another, to uplift one of the most economically challenged regions in the United States. He spoke passionately about the potential impact this collaboration could have, not just in Appalachia but as a model for other regions facing similar challenges. As I listened to him articulate his vision, I immediately knew that Mayor Patterson was the perfect person to write the foreword for "The Entrepreneurial City". Steve embodies the mindset of an entrepreneur, bringing innovation and strategic thinking to both his city and the many organizations he serves. His ability to think creatively and connect different ideas to create a greater whole is truly inspiring. Recently, on the Amazing Cities and Towns Podcast, we discussed his recent trip to Ukraine, where he met with mayors and city officials to explore how his city could share ideas and resources to support a nation in need. Mayor Patterson is a leader who believes in building coalitions and fostering collaboration to ensure mutual success. I am deeply grateful to my good friend and fellow Appalachian for his insightful contributions to this book and for his ongoing commitment to making a positive impact.

MAYOR VINCE WILLIAMS of Union City, Georgia:

Vince Williams is the twentieth Mayor of Union City, Georgia, first elected to the office in 2013 after serving six years on the City Council. Mayor Williams assumed leadership during a period of financial challenges, a stagnant economy, and a community in need of a stable vision. Demonstrating his entrepreneurial spirit, he spearheaded the transformation of a dilapidated mall into a thriving, multi-million-dollar film studio, while also driving the largest job creation initiative in the City's history. Despite initial skepticism, his risk-taking ability and innovative vision turned doubters into believers. Mayor Williams has also been a strong advocate for regional cooperation, working to build consensus and partnerships among South Fulton, Metro Atlanta, and Georgia's various governments, businesses, and civic communities. Under his leadership, Union City was recognized as Georgia's Fastest Growing City in 2023. In 2022, Mayor Williams was elected President of the National League of Cities, where he continues to bring his entrepreneurial spirit to a national and international audience.

JOHN HICKENLOOPER, former Mayor of Denver, former Governor of Colorado, and current United States Senator:

I had the opportunity to meet John Hickenlooper several years ago when the National League of Cities was courting the City of Denver to host the National League of Cities Annual Conference. We met at his loft in the Lower Downtown (LoDo) section of Denver, right beside Coors Field, the home of the Colorado Rockies, and I immediately knew he was one of the most interesting people I had ever met. He related the story of opening one of the nation's first brewpubs, a venture that played a key role in revitalizing the LoDo area, and how this experience led him to run for Mayor of Denver. During his campaign, he used a series of creative television ads that separated him from the rest of the field, including one in which he expressed his unhappiness over a recent increase in parking rates by walking the streets to feed parking meters that were near expiring. He has never been a conventional politician, and his entrepreneurial approach has served him well in business as well as in his meteoric rise to become Governor of Colorado and a United States Senator. Hickenlooper's story underscores the value of thinking outside the box, illustrating how entrepreneurial leaders often blur the lines between business innovation and public service to achieve transformative success.

CLARENCE ANTHONY, CEO & Executive Director of the National League of Cities

One of my closest friends in local government is Clarence Anthony, the former Mayor of South Bay, Florida for over 24 years and now the Executive Director of the National League of Cities. I've had the opportunity to travel the world with Clarence as we participated in numerous United Cities and Local Governments (UCLG) conferences in locations like Jeju Island, South Korea, Istanbul, Turkey, Marrakech, Morocco, and Mexico City, Mexico. Throughout his career, Clarence has had an entrepreneurial mindset and brought it to his community of South Bay, Florida as a young mayor. I watched as he served as the founding Treasurer of the UCLG and later became the Interim Manager for UCLG in Madrid, Spain. I think this Robert Browning quote describes Clarence perfectly, "Ah, but a man's reach should exceed his grasp. Or what's a heaven for?" Clarence Anthony has breathed new life into the National League of Cities and taken the organization into a new century of service in 2024. He led NLC through the COVID-19 pandemic and not only survived but prospered when the Federal government delivered billions of dollars in direct, flexible relief through the CARES Act and the historic American Rescue Plan Act. This was followed by the Infrastructure Investment and Jobs Act which built needed roads, bridges, broadband, water, and sewer to every city in America. Clarence's innovative leadership and forward-thinking approach have not only guided the National League of Cities through challenging times but also positioned it as a catalyst for entrepreneurial growth and resilience in communities across America."

COUNCILMEMBER CONRAD LEE, of Bellevue, WA:

Conrad Lee, a close friend in the local government world, has been a pioneering force in Bellevue, Washington since he first joined the City Council as its first minority member in 1967. Born in China and raised in Hong Kong, Conrad brought a global perspective to Bellevue that has deeply influenced his approach to governance. His greatest strength lies in his relentless pursuit of operational excellence and his ability to forge public-private partnerships that have driven economic growth and community development in Bellevue.

During his tenure as Regional Administrator for the Small Business Administration, Conrad transformed the agency from a distant bureaucracy into a valuable partner for local entrepreneurs. His efforts ensured that resources were accessible to all, particularly minorities, women, and disadvantaged groups, fostering a more inclusive business environment in Bellevue. In my work as a consultant to local governments, I have visited Bellevue on numerous occasions, and Conrad has always been generous with his time and wisdom. Our discussions have not only benefited his city but have also shaped my approach to fostering innovation and collaboration in local government. Through his dedication and visionary leadership, Conrad Lee has not only strengthened Bellevue's community and economy but has also set a standard for entrepreneurial governance that continues to inspire leaders both locally and nationally.

COUNCILMEMBER JON GRIFFITH of Bridgeport, WV:

Jon Griffith, a council member in Bridgeport, WV, is a former teacher and coach who has excelled at marketing his community and being a positive voice on social media. I met Jon when he was first elected to the City Council at a Municipal League meeting and found out that he was one of my daughter's teachers in junior high school. He is a talented writer, and his weekly column is one of my "must reads" each week. He has an eye for detail and does a great job explaining the activities of his city and how they will impact the public.

I'm often asked how council members can make an impact in their city without looking like they are crowding into the territory of the mayor or city manager. Jon has found that balance and is a good example for anyone who would like to see how it's done. Council members can have a significant influence by focusing on communication, public engagement, and transparency. Jon's approach demonstrates that by leveraging their unique backgrounds and skills, council members can become trusted voices in their communities without overstepping their roles. Whether it's writing a regular column, as Jon does, or simply being a visible and accessible presence on social media, council members can build trust and keep residents informed about the decisions being made on their behalf. Jon's work in Bridgeport is a prime example of how to serve as an ambassador for your city and its residents, using creative and thoughtful outreach to make a lasting impact.

MAYOR DERRICK R. WOOD of Dumfries, VA, and entrepreneur:

Derrick R. Wood is a multifaceted leader—a successful author, veteran, entrepreneur, family man, and the current Mayor of Dumfries, Virginia. I first met Mayor Wood when he contacted me to facilitate a retreat for the Dumfries City Council. Demonstrating his entrepreneurial spirit and promotional savvy, he arranged for the retreat to take place at the National League of Cities' new offices in Washington, DC—a strategic move that reflects his commitment to elevating his town's profile.

Despite Dumfries having a population of just under 6,000 residents, Mayor Wood has consistently leveraged his entrepreneurial mindset to put the town on the map. In addition to his role as mayor, Derrick is the founder and pitmaster of Dyvine Barbecue In Motion, a mobile barbecue catering company, and competition team that has gained significant recognition. Over the years, I've had the privilege of watching Mayor Wood grow and evolve into a leader and an inspirational figure within the National League of Cities. His innovative approach and dynamic leadership make him a rising star to watch in local government and beyond.

MAYOR PAUL TENHAKEN of Sioux Falls, South Dakota

While scrolling through TikTok one evening I saw a touching story that epitomizes the concept of an entrepreneurial leader. Mayor Paul TenHaken of Sioux Falls, South Dakota received a video of a bike tunnel covered with graffiti, and the youthful offender was identified and would soon be charged with vandalism. While some in local government would call a press conference with the police chief and announce the capture of the offender, Mayor TenHaken contacted the young man and met with him to come up with a solution for this transgression. Acknowledging his crime, young Jaden Brunz was fined $700 for the removal of the graffiti but the mayor did something that brought national attention to his city. He commissioned Jaden to do a painting that now hangs in the mayor's office, and he paid him $800, which covered his fine. Young Jaden has now committed to doing 'legal' artwork and Mayor TenHaken, himself a graphic artist, is working with city departments to develop several 'art walls' throughout the community so that artists can show off their talents in a way that brings benefit to the city. This act of kindness is not the only reason for recognizing Mayor TenHaken. He has been named by Entrepreneur Magazine in the "Top Ten Emerging Entrepreneurs" as well as South Dakota's "Young Entrepreneur of the Year" for his business success. He has also received accolades for his fiscal responsibility in his city and its lowest debt per capita in over a decade.

JIM BYARD JR., former Mayor of Prattville, Alabama, and Former Director of the Alabama Department of Economic and Community Affairs

Years ago, at National League of Cities events, I often noticed a young man in a bow tie, standing out among a group of older Alabama mayors. That striking figure was Jim Byard Jr., then the Mayor of Prattville, Alabama, and one of the kindest people I've had the pleasure of meeting. It was no surprise when Governor Robert Bentley tapped Jim to lead the Alabama Department of Economic and Community Affairs, where he played a key role in revitalizing downtowns and driving economic development across the state.

Today, as head of Byard Associates, Jim continues to apply his wealth of experience and entrepreneurial mindset in local government to help cities and towns thrive. His expertise in business development and community growth has made him a trusted advisor for fostering entrepreneurial success and empowering communities across Alabama and beyond.

MAYOR VICTORIA WOODARDS of Tacoma, Washington, and Past President of the National League of Cities:

Being an entrepreneurial leader requires energy, drive, and a willingness to roll up your sleeves and outwork everyone else. Mayor Victoria Woodards of Tacoma, Washington, exemplifies these qualities, having established herself as a national leader in housing, equity, opportunity, and youth engagement. Over the years, I've had the privilege of getting to know Victoria through the National League of Cities, and I've found her to be a leader who deeply understands the importance of building coalitions and forming partnerships to benefit her citizens.

During a visit to Tacoma for the NLC Summer Board meeting in 2023, I was struck by the remarkable transformation the city has undergone over the past twenty years—a testament to Victoria's visionary leadership. I was equally impressed by the dozens of Tacoma business leaders in attendance at the board meeting. Many of them emphasized the mayor's ability to collaborate with the business community, a key factor that has driven numerous innovative projects throughout the city. She maintains an open-door policy, actively listening to her constituents and staying informed on emerging trends in local government. Her entrepreneurial spirit is perfectly captured by the theme of her presidential year at the National League of Cities: "People + Partnerships = Possibilities."

VICE-MAYOR JOLIEN CARABALLO of Port St. Lucie, Florida, and Past President of the Florida League of Cities:

I first met Vice-Mayor Jolien Caraballo while conducting leadership training for the City of Port St. Lucie, Florida in 2023. Her passion for her city is evident in her relentless pursuit of innovative strategies to enhance community engagement and strategic planning. Port St. Lucie is one of the fastest-growing cities in the United States, with its population expected to continue rising as more businesses and individuals are drawn to its high quality of life and business-friendly environment.

While Vice-Mayor Caraballo is committed to preserving the "small town feel" that defines her community, she also understands the challenges that come with rapid growth. Without careful planning, the city could easily be overwhelmed by unfocused development. Refusing to take a passive approach, Jolien dedicates significant time and effort to connecting with state and national leaders, securing the resources and funding needed to manage this unprecedented growth effectively.

As a member of the National League of Cities Housing Accelerator Task Force, she collaborates with experts from the American Planning Association to develop national solutions for attainable housing. Vice-Mayor Caraballo's entrepreneurial mindset is driving transformative changes in local government planning, and the positive impact of her efforts is already being felt throughout her city.

RAP HANKINS, Former Councilman in Trotwood, Ohio, and current President of Drive Electric Dayton:

During my time at the National League of Cities, I encountered no one more passionate about their city than Councilman Rap Hankins of Trotwood, Ohio. I first met Rap and his wife, Jan, at an NLC Leadership Training Council meeting, and we have remained close friends ever since. After an impactful career as a city councilman, Rap continued to seek ways to uplift his community through technology and innovation.

As an independent computer technology rep, Rap was acutely aware of the ongoing technology revolution and was concerned that communities like Trotwood might be left behind on the information superhighway. When electric vehicles (EVs) began gaining traction, Rap noticed a stark disparity—few EVs were present in his community, and even fewer EV chargers were available. Determined to address this inequity, Rap joined Drive Electric Dayton (DED) and attended a DED meeting in a large garage. Reflecting on his experience there, he said, "I looked at all the EVs and I looked at the people who owned the cars and realized they were a little different than me. But I thought to myself, if you can do a ride and drive in a wealthy white community, why couldn't you do one in an African American community, too?"

Rap's commitment to bridging this gap led to his election as President of Drive Electric Dayton, where he has since become a national advocate for equity in the EV sector and the expansion of charging infrastructure in communities of color. His efforts exemplify entrepreneurial leadership at its finest using one's skills to identify and meet the needs of underserved communities, thereby making a lasting impact.

CHAD EDWARDS, Village Manager of East Palestine, Ohio:

I first became acquainted with Chad Edwards during his tenure as city manager of Shinnston, West Virginia, a small community in central West Virginia facing the common challenges of declining downtowns and aging infrastructure, with few new businesses or residents on the horizon. With little more than an entrepreneurial mindset, Chad initiated small but impactful improvements that sparked enthusiasm and excitement within the community. A new logo, along with updated street signage, gave the city a fresh, modern look that not only caught the attention of existing businesses but also attracted entrepreneurs considering Shinnston as a potential location.

Despite his busy schedule, Chad made it a priority to build relationships with state and national officials, ultimately earning him a seat at the table where key decisions were being made. Given his talent and dedication, it was no surprise when larger cities began to take notice, leading to his eventual departure from Shinnston. However, what did surprise me was his decision to take on the role of village manager for East Palestine, Ohio—a community still reeling from the February 3, 2023, train derailment and subsequent chemical release.

Chad's move to East Palestine is a testament to his commitment to tackling tough challenges, and I'm eager to see how he applies his skills to what many might consider an impossible task.

MAYOR STEVE WILLIAMS of Huntington, West Virginia:

Imagine being the mayor of a city at the epicenter of the nationwide opioid epidemic. On August 15, 2016, Mayor Steve Williams of Huntington, West Virginia, received the devastating news that 29 people had overdosed in his city in a single day. The city's emergency services worked around the clock to save lives and spread the urgent message that the community was facing a deadly crisis. While many might have crumbled under the weight of such a challenge, Mayor Williams chose to confront it head-on, rallying his community around the bold directive to "Make No Small Plans."

Understanding that drug abuse was only the most immediate symptom of deeper issues, Mayor Williams recognized that the lack of jobs and deteriorating neighborhoods were fueling the demand for illegal drugs. He collaborated with neighborhood leaders to devise plans for revitalizing aging neighborhoods and creating resources to support both youth and impacted adults. Through his relentless pursuit of solutions, Williams explored every possible opportunity to aid his city.

Under his leadership, Huntington was named the $3 million grand prize winner of the America's Best Communities competition in April 2017, for a comprehensive plan to transform the city into the economic gateway of the Appalachian region. This nationwide contest, sponsored by Frontier Communications, aimed to spur economic development in small communities, with more than 350 communities participating. Since the competition began in 2015, Williams' administration has leveraged more than $55 million in additional grants, philanthropic contributions, and corporate investments for the neighborhoods of Highlawn, Fairfield, and West Huntington. Mayor Williams stands as a true example of entrepreneurial leadership, demonstrating how bold vision and determined action can lead to transformative change.

DENISE MITCHELL Mayor Pro-Tem of College Park, Maryland:

The word "entrepreneur" comes from the French verb *entreprendre*, meaning "to undertake"—and no one embodies that spirit more than Denise Mitchell of College Park, Maryland. I've had the privilege of getting to know this energetic leader as she rose to national prominence, becoming President of the National Black Caucus of Local Elected Officials (NBC-LEO) and serving on the Board of Directors for the National League of Cities.

Denise's passion for children shines through her role as Area Manager for AlphaBEST Education, Inc., where she helps create a safe, nurturing environment for students in before- and after-school programs. A natural coalition builder, Denise doesn't shy away from obstacles; she meets them head-on with a smile—her trademark—and a solution-oriented approach that advances her community and the organizations she serves.

JOHN BRENNER, former Mayor of York, Pennsylvania, and current Executive Director of the Pennsylvania Municipal League:

I am often asked what the toughest job in local government is, and I always reply, "Being the director of the State Municipal League!" My good friend John Brenner is one of the brightest and most effective executive directors I have met. Task-focused and dedicated, John consistently takes the time to learn about new innovations and technologies that can benefit his members. As a former mayor, John was always on the lookout for ways to keep his city competitive, and he has brought that same innovative spirit to his current role, advocating on behalf of all cities in Pennsylvania.

I had the pleasure of interviewing John on my Amazing Cities and Towns Podcast, and his episode quickly became one of the most popular we've ever done. What I admire most about John is his ability to provide concrete examples of successful cities in his state, and his encouragement for others to visit and learn from those successes.

MAYOR TERRY WILLIAMS of Spencer, West Virginia:

While some may think that entrepreneurial leadership is the domain of the young, Mayor Terry Williams of Spencer, West Virginia, proves that innovative thinking knows no age. Mayor Williams has held office since the age of 24, and now, at 72, he continues to learn and apply entrepreneurial strategies in his rural town of just over 2,000 residents. After 48 years at the helm, he remains adaptive and willing to tackle challenges that many would find too daunting.

As rural communities across the country struggle with the impacts of online shopping and big-box retailers, Spencer has managed to maintain its hometown charm, with clean, well-maintained storefronts and sidewalks. I once visited Mayor Williams after a tragic fire had devastated part of downtown, and I was struck by how quickly he had mobilized efforts to clean up the site and was already brainstorming ideas for the vacant lot's future.

Mayor Williams has also made it a priority to stay visible in the state capital, advocating tirelessly for Spencer's economic development and securing vital grants. He has served as President of the West Virginia Municipal League three times across three different decades, and he frequently engages with legislators and the Governor to ensure his town receives the attention it deserves. Over his tenure, he has secured more than $100 million in grants and is currently working on projects worth over $10 million. There is much that younger officials can learn from this dedicated and likable mayor.

AMBER MILLER-BELCHER: City Clerk of the City of Logan, West Virginia, and entrepreneur

Amber Miller-Belcher, highlighted in the introduction of this book, embodies the spirit of leadership and resilience as she works tirelessly to uplift her Appalachian community in Southern West Virginia. Amber possesses the rare ability to understand what it truly means to lead, gracefully navigating challenges—even when detractors and online critics attempt to derail her efforts.

During my years of elected service in West Virginia, I encountered many individuals who were negative simply for the sake of negativity. These naysayers would often invest their energy in tearing down the visions of those who dared to dream bigger for their communities.

Amber, however, has been a one-woman public relations powerhouse for the City of Logan, infusing the community with a palpable sense of hope and purpose.

When she received the prestigious Rhododendron Award from First Lady Cathy Justice, Amber's acceptance speech resonated with her unyielding commitment to inspire: "I want to be the light that inspires others to build a better life for generations to come and shine around others to set the standards for excellence and prosperity. I hope I live my life in a way that makes others want to do more and work hard—in a fabulous pair of heels."

Amber's passion for fostering confidence in young girls is demonstrated through the Festival Queen Fundraiser, a program she initiated to provide opportunities for girls from all walks of life to participate in Fairs and Festivals, don a crown, and contribute meaningfully to their community. Amber's efforts ensure that future generations can envision a brighter path, one built with perseverance, pride, and poise.

MAYOR MARY M. DENNIS of Live Oak, Texas, and Past President of the Texas Municipal League:

Several years ago, while attending a Texas Municipal League reception, I was engaged in conversation with a group of mayors and council members when a well-dressed lady caught my attention by looking intently at my nametag. Suddenly, she pointed at me and exclaimed, "You wrote the book I'm reading!" She went on to explain that she had my book, "The Amazing City-7 Steps to Creating an Amazing City", on a table on her patio and was thoroughly enjoying it. That lady was Mary Dennis, the Mayor of Live Oak, Texas, and a Past President of the Texas Municipal League. As the mayor of a small but dynamic city, Mayor Dennis has leveraged her entrepreneurial mindset and leadership skills to transform Live Oak into a vibrant and growing community. She understands the value of staying active on both the state and national levels, recognizing that this engagement not only brings back fresh ideas but also opens doors to the funding necessary for building the infrastructure that supports a thriving economy. Her story is an example of the value of associations like the Texas Municipal League and the National League of Cities.

ANNETTE WRIGHT, City Clerk of Clarksburg, West Virginia:

Throughout my long journey in local government, I can honestly say that I would not be where I am today without the friendship and guidance of my hometown city clerk, Annette Wright. In many cities and towns, fostering an entrepreneurial spirit is nearly impossible without the support of a dedicated city clerk. While elected officials and city managers may come and go, it is often the city clerk who provides the stability that keeps a community on course.

I first met Annette in 1996 when she began her career with the City of Clarksburg, and I watched with admiration as she transitioned from a legal secretary to earning her Certified Municipal Clerk certification in 2002 and her Master Municipal Clerk designation in 2018. Her excellence has been recognized by her peers, as she was named 'Clerk of the Year' in 2005-06 by the West Virginia Municipal League. Annette also broke new ground by serving as Interim City Manager from July 2019 to April 2020, becoming the first female to hold the position.

As I travel across the country, meeting city clerks from both large and small municipalities, I've learned that these individuals often have their finger on the pulse of the community, offering an honest assessment of the "state of the city." Annette Wright exemplifies this role perfectly. She has been a mentor to many officials throughout West Virginia, and she will undoubtedly leave a legacy of service, leadership, and kindness to all who have the privilege of knowing her.

Conclusion

The stories of these remarkable leaders illustrate the profound impact that entrepreneurial thinking can have on local government. Whether it's transforming a struggling economy, revitalizing a community, or tackling seemingly insurmountable challenges, these mayors, councilmembers, city managers, and clerks have demonstrated that true leadership is about more than just managing—it's about envisioning a better future and mobilizing others to achieve it.

Through their innovative approaches, each of these leaders has shown that entrepreneurial leadership is not confined to the business world; it is equally vital in the public sector, where creativity, resilience, and collaboration can drive meaningful change. They have navigated complex landscapes, often in the face of adversity, and have emerged as beacons of hope and progress for their communities.

As you reflect on their journeys, consider how the principles of entrepreneurial leadership can be applied in your community. The path forward may not always be clear, and the challenges may be great, but with the right mindset and a commitment to innovation and collaboration, there is no limit to what can be achieved.

In the end, the true measure of an entrepreneurial leader lies not just in the ideas they generate, but in their ability to turn those ideas into reality, inspiring others to join them in the pursuit of a shared vision. The leaders profiled in this chapter have done just that, leaving a legacy of service, innovation, and transformation that will continue to inspire future generations.

CHAPTER 6

BRANDING IN THE ENTREPRENEURIAL CITY

> *"Your brand is your voice—define it boldly, communicate it fearlessly, and watch it transform every connection into a powerful impact."*
> *Jeni Bukolt, Founder of* **HAVEN** *Creative*®

Several years ago, my wife and I visited our daughter in Lancaster, South Carolina. She mentioned that she wanted to take us to a nearby town with a railroad track running along Main Street. If we got there at the right time, she said, we might see the train spewing black smoke as it made its way through town. Intrigued, my wife and I joined our daughter, son-in-law, and our two granddaughters, and headed to Waxhaw, North Carolina.

The town was charming, with its antique shops and clothing stores. We enjoyed dinner at Maxwell's Tavern, a pub-style restaurant with a fun vibe and a great menu. When the train came through town, we jumped out of our seats and rushed to the window to watch it chug along. After dinner, we walked to the local ice cream shop for dessert. I turned to my wife and said, "This is one of the nicest small towns I've ever visited." She asked why, and I told her it just felt like someone had put in a lot of time to create a sense of hospitality and cohesiveness that you don't find in many places.

When I returned home to West Virginia, my son-in-law mentioned that he had met Jeni Bukolt, the person who helped Waxhaw with its branding. He thought she might make a great guest for my podcast. I reached out to Jeni, and she quickly agreed to be on the show. During our conversation, I learned that her company, **HAVEN** *Creative®*, had done a complete rebranding for Waxhaw. This included a website re-design, new welcome signage, wayfinding elements, and many hours working with the community to identify the town's theme and vision. The more I spoke with Jeni, the more I realized this kind of thoughtful branding is the missing piece for many communities that are "almost" there but lack that certain something to be truly amazing!

I've stayed in touch with Jeni and continue to follow the work of her incredible team as they transform communities and businesses from their base in Charlotte, North Carolina. I've included contact information for **HAVEN** *Creative®* in Appendix 1 at the back of this book.

Branding is crucial for cities because it shapes their identity, reputation, and economic future. A well-developed brand can set a city apart by highlighting its unique history, culture, and business climate. While many people believe branding is primarily about boosting tourism, the internet, and social media have made it a valuable tool for economic development. A strong brand can foster community pride and engagement, encouraging residents to promote their community to others. For potential businesses and future residents, their first encounter with a community is often through its online presence. A strong, consistent brand can make a lasting impression and pay dividends for years to come.

What is the difference between corporate branding and city branding?

✦ Most people are aware of corporate brands and their value to the companies that own them. Major brands like McDonald's, Exxon, Apple, and Amazon use and protect their brand to attract customers, increase market share, and build brand loyalty. They spend millions of dollars each year to drive sales, enhance brand recognition, and differentiate themselves from competitors. They want the consumer to recognize their brand when they are driving down a highway or seeing a truck pull up to their homes.

✦ City branding differs in both objective and purpose. The objective of a city brand is to enhance the city's reputation, increase tourism, inspire civic pride, and foster investment and economic development. The Raleigh, Durham, and Chapel Hill region is an example of what branding can bring to a city or region. The brand of "The Research Triangle" has resulted in billions of dollars in investment and worldwide recognition as a place to grow and develop world-class companies. The purpose of the branding is to promote economic growth, community, and cultural engagement.

Figure 3 When almost every visitor to your city has a smartphone camera in their pocket, it is easy to identify your brand with potential visitors and future residents.

Components of a City Brand

✦ **Visual Identity**- Cities and towns have a myriad of logos, colors, and designs that need to be researched and coordinated when building an effective brand. Buildings, vehicles, equipment, public spaces, websites, clothing, signage, etc., are all important in sending the right message. Providing acceptable uses of the brand and ensuring that each department in a city understands the proper display, colors, and designs are critical to sending a consistent message.

✦ **Slogans and taglines**- While most people recognize that "Keep Austin Weird" has become a defining slogan for Austin, Texas, do most people connect "Land of the Sky" with Asheville, North Carolina? It takes time to develop creative slogans and taglines for a city or region. It is important to see what is currently being used and determine whether revising or changing a well-known slogan or tagline will serve the city's needs. The audience for the slogan or tagline should be taken into consideration. Sedona, Arizona's tagline of "The Most Beautiful Place on Earth" might be an accurate description in the minds of people who live there, but will it achieve the purpose of enticing visitors to the city?

✦ **Key Messages and Narratives**- When developing a city's brand, key messages and narratives should be thoughtfully crafted to reflect the city's unique identity, strengths, and aspirations.

1. Unique Identity and Character

2. Economic Opportunities

3. Quality of Life

4. Innovation and Sustainability

5. Community and Inclusivity

6. Tourism and Attractions

7. Education and Research

8. Future Vision

1. Examples of Successfully Branded Cities and Why

    1. **Nashville, Tennessee:** "Music City" capitalizes on the rich musical history and its position as the heart of country music. Venues like the Grand Ole Opry and The Bluebird Café draw locals and tourists alike, attracting millions to the city annually. With a well-known brand and attractive business climate, Nashville boasts a booming economic climate for the city and surrounding areas.

    2. **Bend, Oregon:** The "Outdoor Playground of the West" highlights the skiing, hiking, mountain biking, and river activities. A great quality of life and numerous craft breweries appeal to residents and visitors alike. The brand promotes a healthy lifestyle and natural beauty. In my only visit to Bend, Oregon, I

can attest to the outdoor vibe in the community. Residents and visitors alike seem dressed for a hike with flannel shirts and Timberland boots.

3. **Hershey, Pennsylvania:** "The Sweetest Place on Earth" highlights the city's identity as the home of Hershey's Chocolate and gives the tourist attraction a visual picture of what awaits visitors at attractions like Hershey Park, Hershey's Chocolate World, and the Hershey Story Museum. It also emphasizes its family-friendly attractions and entertainment options.

4. **Huntsville, Alabama:** "Rocket City" identifies Huntsville for its significant role in the U.S. space program and the presence of NASA's Marshall Space Flight Center. It is also pivotal in attracting innovation and aerospace companies to the city and region. I can remember visiting Huntsville for a speech to the Alabama League of Municipalities and seeing a life-size statue of a NASA astronaut in the lobby of the hotel.

5. **Orlando, Florida:** "The Theme Park Capital of the World" has effectively worked for this Florida city. It transmits the message that Orlando is more than Walt Disney World and offers a destination for family vacations and entertainment. It has also gained worldwide attention for the city and brings in millions of international visitors each year.

6. **Tulsa, Oklahoma:** "A New Kind of Energy" highlights an evolving economy, particularly in the energy, aerospace, and technology sectors. It also sends a message about a high quality of life with the world-class, Gathering Place riverfront park and venues like the Tulsa Arts District and the Woody Guthrie Center.

Developing a Strong City Brand

Before a city can build a strong brand, it must first gain a clear understanding of its current perceptions and identity in the eyes of key stakeholders—citizens, visitors, the business community, and others. This process begins with gathering data on these perceptions through interviews, surveys, focus groups, and other methods. These efforts will help identify the city's strengths, weaknesses, and unique characteristics that could impact the brand.

In addition to gathering perceptions, it's essential to assemble and review any existing materials—both printed and digital—to assess how the city is currently communicating its identity. This audit will reveal whether the city's branding is consistent and aligned with its desired image. It may also highlight areas where the branding needs updating or even indicate a need for a complete rebranding effort.

A key decision in this process is whether to engage the services of a professional branding organization or manage the branding effort in-house. This decision is crucial, as a city's brand is a long-term investment that can influence public perception and city growth for years to come. If the branding process is not handled correctly from the outset, the cost of redoing the work later could far exceed the initial investment. Engaging a professional may be advisable if there are concerns about the capacity or expertise available within the city's team.

One valuable lesson I've learned from my time in local government is that the process of selecting a branding firm can be an educational experience in itself. Sitting in on presentations from companies vying for your city's branding project can provide valuable insights into current trends, innovative strategies, and the intricacies of branding that you might not encounter in your day-to-day work.

Since branding is still a relatively new endeavor for many cities, understanding what is involved and what to expect from branding firms can better equip you to make informed decisions and ensure a successful outcome.

Defining the Brand Goals

Once you have a clear understanding of the city's current brand and the perceptions associated with it, the next critical step is to define the brand goals. Setting clear and achievable goals is essential to ensure that the branding efforts align with both the city's short-term needs and long-term vision.

Building a strong brand is a process that can span several months to several years, and it's important to recognize that regular monitoring and updating will be necessary to keep the brand relevant and effective. By understanding this, you can develop goals that are not only strategic and forward-looking but also flexible enough to adapt to changing circumstances and evolving community needs.

Short-term Goals: Short-term branding goals often focus on immediate and visible items that can gather support and build enthusiasm. These goals might include:

✦ Increasing Awareness: If your city is relatively unknown or has a poor reputation, a short-term goal might be to raise awareness about the upcoming new branding campaign.

✦ Enhancing Visual Identity: Updating the visual elements in the city, such as the city's logo, color schemes, and visual elements can be a boost that provides excitement to residents and visitors.

✦ Revitalizing Communication Channels: Unveiling a new website or social media design can begin the effort to bring consistency across all communication mediums.

✦ Engaging the Community: A brand unveiling or brainstorming workshop can give community members pride and ownership in the new brand.

Long-term Goals: Long-term goals are more strategic and focus on developing a brand that will grow alongside the city's growth and needs. These might include:

✦ Building a Strong Reputation: Cultivating a reputation for quality of life, economic opportunities or cultural vibrancy is a long-term effort that involves consistent messaging, community engagement, and strategic partnerships.

✦ Promoting Community Pride: A good city brand can help promote a sense of pride and belonging among residents, which in turn supports civic engagement, volunteerism, and a stronger community fabric.

✦ Attracting Economic Development: By developing a brand that highlights the city's strengths—such as its business-friendly environment, skilled workforce, or prime location—you can attract long-term investment and encourage sustainable economic development.

Implementing the Brand Strategy:

Effective implementation of the brand strategy requires careful planning and coordination across all city departments. The new brand should become an integral part of the city's culture as quickly as possible to ensure consistency and avoid costly mistakes. For example, imagine investing thousands of dollars in developing a new brand, only to discover that the Public Works Department has just ordered twenty new dump trucks, all featuring the old logos and color schemes. To prevent such scenarios, it's essential to establish clear guidelines and communication channels.

Coordinating Across Departments:

Distribute detailed instructions to all city departments to ensure that the new brand is consistently applied across all materials, equipment, and communications. This includes everything from vehicle decals and uniforms to official documents and digital assets. City departments should be informed of the changes well in advance to avoid any missteps during the transition.

Engaging Employees:

Encourage employees to embrace the new brand by offering incentives such as trading in old uniforms or branded items for new ones featuring the updated logo. This not only ensures brand consistency but also boosts morale and enthusiasm for the rebranding effort. Employees who feel involved in the process are more likely to become advocates for the new brand.

Collaborating with Local Businesses:

Local businesses play a crucial role in promoting the new city brand. Work with them to share the story behind the rebranding and explore creative ways to get them involved. For instance, distributing coffee mugs with the new logo at a Chamber of Commerce event or hanging a banner across Main Street can generate excitement and word-of-mouth promotion. Consider organizing events or offering promotional materials that encourage businesses to align their own branding with the city's new identity.

Ensuring Consistent Messaging:

Finally, make sure that all external communications are aligned with the new brand strategy. Provide guidelines to media outlets and any organizations that use the city brand in advertisements, brochures, and other materials. Consistency is key to building recognition and trust in the new brand, so every touchpoint should reinforce the city's updated identity.

In conclusion, branding is one of the most powerful tools available to an entrepreneurial city. However, branding efforts must be professionally executed, inclusive, and comprehensive. Entrusting your city's brand to someone without the necessary expertise—like a cousin or a brother-in-law—can risk damaging the goodwill that your city has built. A strong, consistent, and authentic brand is a key to fostering confidence and growth in your community.

Footnote: If you are in doubt about creating an impactful brand for your city, take Nike's advice: *Just Do It!*

CHAPTER 7
SUCCESSFUL EXAMPLES OF CITY BRANDING

> *"You can't rely on bringing people downtown, you have to put them there."*
> **Jane Jacobs, The Death and Life of Great American Cities**

Pittsburgh, Pennsylvania:

I have had the opportunity to observe firsthand the remarkable transformation of Pittsburgh from the "Steel City" to a thriving "Tech Hub." Living just an hour and a half south of Pittsburgh, I witnessed the decline of the steel manufacturing industry and the resulting economic devastation that left the city grappling with high unemployment and a struggling economy. For years, rusting factories dominated the downtown riverfront, and failing infrastructure was a common sight across the city.

Pittsburgh's rebranding journey began with a strategic focus on education, healthcare, and technology. Leveraging the presence of prestigious institutions like Carnegie Mellon University and the University of Pittsburgh, the city attracted the attention of tech companies eager to tap into local talent and innovation. This focus on technology and research led to the gradual growth of Pittsburgh as a center for robotics, artificial intelligence, and biomedical research.

Figure 4 The Three Rivers converge in Pittsburgh, PA

Today, Pittsburgh is celebrated as a leader in technology and innovation. The city's rebranding has not only revitalized its economy but also sparked renewed enthusiasm and energy, drawing in a growing population of young professionals who are excited to be part of its future.

Chattanooga, Tennessee:

Once known as the "Dynamo of Dixie," Chattanooga experienced a similar decline to that of Pittsburgh as its massive foundries and factories, which once produced thousands of jobs, also contributed to severe pollution. As the industrial sector contracted, the city faced significant economic hardships and population loss. While manufacturing remains an important part of the local economy, Chattanooga recognized the need for a new identity to secure its future.

Figure 5 The Welcome Sign to in the Chattanooga Airport

In 2010, Chattanooga became the first city in the United States to roll out a citywide gigabit network, a bold move that marked the beginning of its transformation into "Gig City." This initiative was spearheaded by an innovative mayor who, having served in the state legislature when the concept of Gig City was born, saw the potential in building out a fiber optic network. This network was initially intended to support a smart grid for the city, but it quickly became the foundation for a much broader technological renaissance.

As Mayor Berke explained in an article for *PCMag*, "The fiber changed our conception of ourselves. The day before we turned it on, we had no belief we could become a tech city. The day afterward, when we became the city with the fastest, cheapest, most pervasive internet in the world, we started a mission to figure out how to use it."

Today, Chattanooga is capitalizing on its growing technology sector, cleaner manufacturing practices, and strategic geographic location. Coupled with its affordable cost of living, the city has redefined itself as a dynamic and growing hub of innovation, attracting new residents, businesses, and opportunities.

Rock Hill, South Carolina:

Rock Hill was once a thriving hub of the textile industry, with many well-known manufacturers calling the city home. However, as the industry restructured and began moving jobs overseas, Rock Hill faced economic decline and rising unemployment rates. Adding to the city's complex history, Rock Hill was the site of two significant events in the civil rights movement: a sit-in at the segregated McCrory's lunch counter and the brutal beating of civil rights leader John Lewis by a white mob—an event that drew national attention.

Figure 6 Downtown Rock Hill, South Carolina

Recognizing the need to transform its image, the city, under the leadership of former Mayor Doug Echols, embarked on a journey of rebranding and revitalization. Mayor Echols, who served on the board of the National League of Cities with me, played a pivotal role in these efforts. In 2002, John Lewis returned to Rock Hill, invited as a speaker at Winthrop University. During this visit, Mayor Echols presented Lewis with the Key to the City and issued an official apology on behalf of the city for the Freedom Riders' treatment in 1961. This moment was a significant step in reconciling the city's past with its vision for a more inclusive future.

Rock Hill was also successful in transforming a donated industrial site into a world-class BMX Supercross Track, which hosted the 2017 BMX World Championships and continues to attract both national and international events. The city's proximity to Charlotte, North Carolina, has further bolstered its growth, providing access to the Charlotte International Airport and drawing a steady stream of visitors who train and compete at the Supercross Track and the Rock Hill Velodrome, an outdoor facility that meets Olympic standards.

During a visit to Rock Hill several years ago, I met with Mayor Echols and asked him to share the planning process that led to the city's emergence as a sports tourism destination. He took me to the City Manager's office, where a bookcase filled with strategic planning documents was neatly arranged. Each step of the city's transformation had been carefully laid out, beginning when Rock Hill was grappling with its identity and vision. This meticulous planning and forward-thinking approach were key to Rock Hill's successful rebranding and revitalization.

Common Pitfalls and How to Avoid Them

Branding is a powerful tool and as the examples above indicate, there is much to be gained by implementing a successful rebranding, but it is wise to note that branding is a complex process that can be fraught with challenges. Understanding the common pitfalls and learning from real-life examples of branding failures can help your city navigate this complex process more efficiently

1. **Failing to Solicit Input from Stakeholders:**

   - **Pitfall:** One of the biggest mistakes in rebranding is not involving key stakeholders—residents, businesses, and community leaders—from the beginning. When stakeholders feel excluded from the process, they are more likely to resist change or criticize the rebranding effort because they lack a sense of ownership.

   - **How to Avoid It:** To prevent this, involve stakeholders from the outset. Engage them through surveys, public forums, and focus groups to gather their input and ensure they feel invested in the outcome. This approach not only fosters buy-in but also enriches the rebranding process with diverse perspectives.

2. **Inconsistency in Messaging:**

   - **Pitfall:** Inconsistent application of the new brand across different city departments and platforms can undermine the rebranding effort. If certain departments, like the water department, are overlooked during the branding process, it can lead to outdated logos, inconsistent messaging, and a significant drop in morale.

   - **How to Avoid It:** To avoid this pitfall, conduct a thorough audit of all branding platforms and ensure that every department is included in the rebranding process. Clear communication and comprehensive guidelines are essential to maintain consistency across the city's messaging and visual identity.

3. **Failing to Be Authentic:**

   - **Pitfall:** Rebranding efforts that attempt to completely reinvent a city without considering its history and cultural identity often fall flat. Ignoring the city's past can alienate long-time residents and create a disconnect between the brand and the community.

- **How to Avoid It:** A successful rebranding should acknowledge and honor the city's heritage while also presenting a vision for the future. The new brand should create a bridge between where the city has been and where it is headed, ensuring that the brand feels authentic and resonates with both residents and visitors.

The Savannah Bananas & Other Unexpected Branding Opportunities

Branding often presents unexpected opportunities for cities to enhance their image. A prime example is the Savannah Bananas, a minor league baseball team in Savannah, Georgia that transformed into a unique brand under the leadership of its colorful owner, who invested his life savings into the team. To attract fans, he introduced creative elements like dancing players and an all-inclusive ticketing system, turning games into a must-see entertainment experience. This unplanned branding success has significantly boosted Savannah's modern appeal, complementing its historic charm.

Similarly, the city of Normal, Illinois, has a unique name which my good friend, Sonja Reece, a former Normal Council member leveraged at every opportunity. She would show up at every National League of Cities Conference with her late husband, Jerry, and announce to everyone that they were the only two Normal people at the conference. By embracing the uniqueness of being "Normal," the city has created a positive and engaging image that resonates with both residents and visitors. Ideas like displaying large letters spelling "Normal" in public spaces have the potential to become social media hotspots, further enhancing the city's brand.

Lastly, Albuquerque, New Mexico, benefited from the popularity of the television series *Breaking Bad*, television series, which brought fans to the city to visit iconic locations from the show. Places like the Tip Top Car Wash regularly see visitors taking selfies out front and having their cars washed. Embracing this unexpected branding opportunity has made Albuquerque a trendy destination for visitors, showing how cities can leverage pop culture to enhance their brand.

These examples demonstrate that branding can sometimes be opportunistic and, when embraced, can significantly enhance a community's image and appeal.

Conclusion:

The Future of Branding in Entrepreneurial Cities

As we conclude this chapter on entrepreneurial city branding, it's clear that the future of branding in cities is poised to be a growing trend. Cities are increasingly recognizing the immense value of a strong brand, and the methods of implementing and maintaining these brands have become more accessible than ever before. Advances in technology, such as digital printers and die-cut machines, have significantly reduced the costs associated with branding. Where once the expense of branding elements—like vehicle logos and city signage— could be prohibitive, these tools now make it far more affordable and straightforward. This ease of access allows cities to make modifications or tweaks to their brand more easily than they could even five to ten years ago.

On the digital front, social media, websites, and online communications like newsletters have revolutionized the way cities manage and update their brands. The simplicity of enacting a brand change or update in the digital realm means that more cities will likely embrace branding as a key strategy for growth and identity.

However, it's important to strike a balance between agility and stability in branding. While it's easier now to make changes, cities should avoid altering their brands too frequently, as this can lead to confusion and a loss of trust. A prime example is the stadium in Miami, originally named Joe Robbie Stadium. Over the years, it has undergone multiple name changes—Pro Player Stadium, Dolphins Stadium, Land Shark Stadium, Sun Life Stadium, and now Hard Rock Stadium. These frequent changes not only confused fans but also diluted the stadium's identity. If there was such a thing as branding malpractice, the Miami stadium might be the all-time worst offender.

In contrast, look at the branding success of Key West, Florida. This Southernmost city has cultivated a unique and trusted brand identity, known for its vibrant tourism, high quality of life for residents, and a distinctive atmosphere as "The Conch Republic." This brand resonates with both tourists and locals because it is authentic and consistent, conveying a clear message of independence and fun. It illustrates how a well-managed brand can become a trusted asset that people are willing to invest in, whether as visitors or as business owners.

CHAPTER 8

GENERATIVE AI IN LOCAL GOVERNMENT: OPPORTUNITIES AND CHALLENGES

Artificial Intelligence is neither good nor evil. It's a tool.
It's a technology for us to use.
Oren Etzioni

I vividly recall the first time I encountered the term "Artificial Intelligence." I wondered whether cities and towns would spearhead this technological revolution or be left in its wake. The potential for AI to transform every facet of local government operations was immediately apparent. From traffic and energy management to public safety, the ability to analyze vast amounts of data in mere seconds promised significant improvements in response times and efficiency. As we delve deeper into AI, its applications in cities and towns appear boundless, limited only by our ability to integrate it into existing and future systems.

A few years ago, I had the opportunity to visit Franz Loewenherz, the Mobility Planning and Solutions Manager for the City of Bellevue,

Washington. The innovative research underway in this forward-thinking city was truly inspiring. Bellevue was pioneering predictive safety measures to protect pedestrians and integrating cutting-edge mobility technologies to enhance its public transportation network. This experience highlighted the potential for local governments to lead in world-class innovation. It also underscored the importance of collaboration between local governments and academic institutions, with the University of Washington playing a pivotal role in this groundbreaking research.

An October 18, 2023, report by Bloomberg Philanthropies titled "State of the Cities: Generative AI in Local Governments" revealed that 96% of surveyed mayors expressed interest in using Generative AI. However, only 2% of cities are actively implementing this technology, while 69% are exploring or testing it. This optimistic outlook for the future of AI in cities and towns underscores the significant need for education and exploration to maximize its potential. The report also identified "lack of awareness" and "budget constraints" as major barriers to AI adoption. This presents forward-looking cities and towns with a unique opportunity to innovate and harness AI's capabilities. Organizations like the National League of Cities, the U.S. Conference of Mayors, and State Municipal Leagues can play a crucial role in providing the training and education necessary to advance this transformative technology.

Inherent Advantages of Using Generative AI in Local Government Operations

Data-Rich Environment: Cities and towns generate vast amounts of data daily, from traffic flow and utility usage to public service requests and social media interactions. Generative AI can quickly analyze and interpret this data, providing insights and recommendations that can optimize operations and enhance efficiencies. For example, AI algorithms can identify patterns in energy consumption, suggest ways to reduce waste, and even predict maintenance needs for public infrastructure before issues arise. The ability to process and utilize large data sets in real-time is a game-changer for local governments looking to improve service delivery and operational efficiency.

A Multitude of Repeatable Patterns: Many city operations involve routine and repeatable tasks, such as traffic management, waste collection, and building maintenance. Generative AI excels in recognizing and optimizing these patterns. For instance, AI can analyze traffic data to adjust signal timings dynamically, reducing congestion and improving traffic flow. In waste management, AI can optimize collection routes, leading to cost savings and more efficient resource use. Additionally, city boundaries provide a well-defined "geo-fenced" area, allowing AI tools to produce highly accurate and localized strategies. This focused application of AI ensures that the technology addresses the specific needs and challenges of each city or town.

Engaged Populations: Generative AI can significantly enhance citizen engagement by enabling local governments to make more informed, data-driven decisions. AI tools can analyze public feedback, social media trends, and survey results to provide insights into public opinion and priorities. This helps city officials understand community needs and respond more effectively. Furthermore, AI can facilitate participatory budgeting processes and other forms of citizen engagement, making it easier for residents to have a say in how resources are allocated, and policies are shaped. By fostering a more inclusive and responsive governance model, AI helps build trust and collaboration between local governments and their communities.

Predictive Capabilities: Generative AI's predictive capabilities can help cities anticipate and prepare for future challenges. For example, AI can forecast demand for public services based on demographic trends, predict the impact of environmental changes, and model the potential outcomes of different policy decisions. This proactive approach allows local governments to address issues before they become critical, improving resilience and sustainability. By leveraging AI's predictive power, cities and towns can plan more effectively and allocate resources more strategically, leading to better outcomes for residents.

Enhanced Public Safety: AI can play a crucial role in enhancing public safety by providing real-time monitoring and analysis of various data streams. For instance, AI-powered surveillance systems can detect unusual activities and alert authorities to potential threats. Predictive policing models can help law enforcement anticipate and prevent criminal activity. In emergency response situations, AI can analyze data from multiple sources to coordinate resources and provide first responders with critical information. These capabilities contribute to a safer and more secure urban environment.

Efficiency and Cost Savings: Implementing Generative AI in city operations can lead to significant efficiency gains and cost savings. Automation of routine tasks frees up human resources for more complex and strategic work. AI-driven optimization of processes reduces waste and minimizes the need for corrective actions. For example, predictive maintenance of infrastructure can extend the lifespan of assets and reduce repair costs. By improving efficiency and reducing operational costs, AI allows local governments to allocate resources more effectively and deliver better services to their constituents.

In conclusion, the inherent advantages of using Generative AI in local government operations are vast and transformative. From optimizing routine tasks and enhancing public safety to improving citizen engagement and predicting future needs, AI has the potential to revolutionize how cities and towns operate. By embracing this technology, local governments can become more efficient, responsive, and innovative, ultimately creating smarter and more resilient communities.

A Word of Caution for the Use of Gen AI in Local Governments: While Generative AI offers immense potential to revolutionize local government operations, it is not without its risks and challenges. One of the primary concerns is the possibility of biased or inaccurate outputs, which can arise from the data used to train AI models. If not carefully managed, these biases could reinforce existing inequalities or lead to misguided decisions that affect public policy and resource allocation. Additionally, the integration of AI into government systems raises significant privacy and security issues, as sensitive citizen data could be vulnerable to breaches or misuse. The opaque nature of AI decision-making, often referred to as the "black box" problem, can also erode public trust if citizens and officials do not fully understand how AI-driven conclusions are reached. Moreover, the rapid advancement of AI technologies could outpace the ability of local governments to regulate and oversee their deployment effectively, leading to ethical dilemmas and unforeseen consequences. Therefore, while embracing Generative AI, local governments must exercise caution, implementing robust oversight, clear ethical guidelines, and continuous evaluation to ensure that these tools are used responsibly and equitably.

CHAPTER 9

ENTREPRENEURIAL PUBLIC SAFETY STRATEGIES

Introduction

One morning, as I was leaving for work, my neighbor stopped me with unsettling news—there had been an attempted break-in at his house the previous night. At around 3:00 AM, a loud noise startled him and his wife from their sleep. Investigating the sound, he found nothing unusual and dismissed it as perhaps a truck backfiring. It wasn't until the next morning that he discovered a boot mark on his side door, a chilling sign that someone had indeed tried to force their way in while they were asleep.

Our peaceful neighborhood suddenly felt less secure. I told my wife to double-check the locks and be extra vigilant at night. Soon after, I noticed my neighbor installing a video doorbell on the same side door where the attempted break-in occurred. He told me the city had a program to provide video doorbells at a reduced price and got one for his side door. He already had one on his front door but realized the need for more comprehensive coverage.

Just a few days later, I saw a police cruiser at his house. The video doorbell had captured footage of two teenagers knocking on his door late at night. The police, suspecting mischief, apprehended the teens, who quickly admitted to the incident. It was a relief to know it wasn't a serious threat, just a couple of teenagers playing pranks.

This experience underscored for me the significance of modern, affordable technology in public safety. The video doorbell quickly and effectively resolved what could have been a much more serious situation. It made me realize that we are living in a new age of policing, where technology not only enhances our sense of security but also plays an active role in preventing and solving crimes.

In an era of rapid technological advancement and evolving societal challenges, public safety cannot rely solely on traditional methods. The entrepreneurial city must adopt innovative strategies that leverage technology, community engagement, and private-sector partnerships to enhance public safety. This chapter explores how entrepreneurial thinking can be applied to public safety, transforming it from a reactive service into a proactive and collaborative effort that ensures the well-being of all citizens.

Redefining Public Safety Through Technology

The integration of technology into public safety operations is a cornerstone of entrepreneurial thinking. Cities across the globe are increasingly using advanced tools such as drones, artificial intelligence, cameras, and data analytics to predict and prevent crime, improve emergency response times, and optimize resource allocation.

Uncrewed Aerial Vehicles (Drones)

One of our clients at Bearing Advisors was a company named BRINC Drones whose mission is to revolutionize public safety by leveraging technology to de-escalate dangerous situations. Each drone deployed to a dangerous situation is one less individual in harm's way, and a potential life saved. We were engaged to introduce their innovative drones to elected officials and explain the concept of "drones as a first responder". The founder of BRINC Drones is an impressive young man named Blake Resnick, who started college at 14 years old and built a fusion reactor in his garage.

Figure 7 BRINC founder and CEO Blake Resnick allows city officials to fly the drone at a meeting in Seattle, WA

He was spurred to action to build technology for first responders in 2017 after the deadliest mass shooting in American history occurred in his native Las Vegas at the Mandalay Bay Hotel. Realizing the challenges first responders faced during the crisis as well as the dangers they encountered daily, Blake saw the need for implementing new technologies into public safety scenarios that could help save lives.

During the time of our work with BRINC Drones, my son, who is a police officer and a member of the Greater Harrison Drug Task Force in Harrison County, West Virginia, told me that he had been selected to undergo FAA training to be licensed to fly drones in his work combatting drug trafficking in our area. Hearing from him how drone technology has made dramatic changes in the safety of drug investigations gave me a real-life experience with this technology.

Another example of the use of drones is the city of Chula Vista, California, which has deployed drones as first responders. These drones arrive at crime scenes or emergencies before officers, providing real-time video feeds that allow for better-informed decision-making. This innovative approach has reduced response times and enhanced the safety of both officers and the public.

In addition to the use of drones in Police Departments, other public safety departments such as Fire and EMS are using the technology and having great results. The ability to hover over fire and accident scenes can transmit critical information to firefighters and alert them of impending danger or finding injured or trapped victims through infrared sensor images.

EMS departments are experimenting with drone delivery of life-saving medications or equipment in difficult-to-access locations. Lightweight payloads can be carried by drones to deliver Naloxone or other overdose medications, and some companies are developing larger drones that can dispatch automated external defibrillators (AED).

An article in the NIH National Heart, Lung and Blood Institute reported on a team of NHLBI-funded researchers that are testing drones to deliver AEDs that can be used to help quickly revive people who've had an out-of-hospital cardiac arrest, a leading cause of death in the United States and worldwide. Each year in the U.S. alone, about 350,000 OHCAs occur, and they are almost always fatal: only about 10% of victims survive.

Quoting the article, "Nicole Redmond, M.D., Ph.D., M.P.H., an NHLBI program officer and chief of the Clinical Applications and Prevention branch in NHLBI's Division of Cardiovascular Sciences, said that in addition to speeding delivery of AEDs to where they are needed most, the drone can do something else: help reduce geographic disparities in the treatment of OHCAs.

"We see AED devices at offices and stadiums in larger urban areas, but not as frequently in rural and remote areas," Redmond said. "But cardiac arrest can happen anywhere. Being able to deploy those devices on-demand via drone to areas where they are lacking could make a big difference."

(Source: NIH National Heart, Lung, and Blood Institute Research Feature, July 24, 2024)

The integration of uncrewed aerial vehicles (drones) into public safety has the potential to revolutionize emergency response across various sectors, including law enforcement, fire services, and emergency medical services (EMS). Companies like BRINC Drones, founded by visionary Blake Resnick, are at the forefront of this innovation, providing first responders with advanced tools that enhance situational awareness and reduce response times. In Chula Vista, California, for example, drones have already demonstrated their value as first responders, delivering real-time video feeds that aid decision-making and protect both officers and civilians. The versatility of drones extends beyond crime scenes; fire departments utilize them to monitor and assess fire scenes, while EMS teams experiment with delivering life-saving medications and equipment to hard-to-reach locations. Notably, researchers funded by the NIH National Heart, Lung, and Blood Institute are exploring the use of drones to deliver automated external defibrillators (AEDs) to out-of-hospital cardiac arrest (OHCA) victims, addressing a critical need for rapid response in rural and remote areas. As highlighted by Dr. Nicole Redmond, drones have the unique capability to bridge the gap in emergency care, ensuring that life-saving devices are available wherever and whenever they are needed, thereby significantly improving outcomes for OHCA victims.

Cameras, Sensors, and Smart City Technology

Like the story of my neighbor's video doorbell episode, the increased clarity and ability to record large amounts of data have put cameras on the front lines of public safety throughout the world. England has developed a CCTV system that covers vast amounts of areas within cities and towns, and it is not uncommon to see criminals regularly identified via the camera system. While many U.S. cities have not developed their systems as thoroughly as England, the use of cameras is growing, supplemented by the huge increase in business and residential security cameras and the millions of "doorbell" cameras that record both delivery thefts and activity within the field of vision for the camera. It is not uncommon for a police department to go into a neighborhood and request video from homeowners to assist in an investigation. Pole-mounted cameras, equipped with license plate readers are also being used in cities and towns and increasingly, private homeowners' associations. While opposed in some areas for privacy concerns, many citizens seem willing to trade privacy for increased security.

With the increasing transporting of hazardous materials on highways and railways, the danger of derailments and accidents is on many public safety official's minds. Some cities and towns have installed sophisticated sensors to monitor environmental conditions to alert city officials to potential public health risks or help in emergency planning for natural disasters, like floods or wildfires. The derailment of a Norfolk Southern freight train carrying hazardous materials in the city of East Palestine, Ohio on February 3, 2023, resulted in hundreds of millions of dollars in cleanup and mitigation expense, in addition to health and environmental concerns for thousands of citizens. This tragic event pointed out the need for improved coordination with emergency officials and the need for advanced warning through detection devices. Sensors play an important role for cities and towns to protect citizens as well as the first responders who often arrive on the scene with no knowledge of the chemicals or toxins they might encounter.

Using Generative AI to Assist Public Safety

As discussed in the previous chapter, Generative AI is a powerful tool and is a welcome addition to public safety departments around the country. The ability to process vast amounts of data can automate duties that, in many cases, are just not being completed in the overworked world of many city and town public safety departments.

With the advent of technology in police, fire, and EMS departments, the production of mountains of data was probably inevitable but daunting, nonetheless. The need to promptly comply with open records requests for body camera footage and other digital information has put a strain on public safety agencies and the need to protect private and confidential information has led to the use of Generative AI to address these needs. There are still concerns and public agencies need to be aware of the capabilities of AI, but its use is almost certain to grow due to staffing and administrative shortages.

Moreover, predictive policing, powered by AI, can analyze vast amounts of data to identify patterns and predict potential crime hotspots. By deploying resources more effectively, cities can prevent incidents before they occur, shifting from a reactive to a proactive public safety model.

Community Engagement and Co-Production of Safety

An entrepreneurial approach to public safety recognizes the critical role of community engagement. Public safety is not solely the responsibility of law enforcement agencies; it is a shared responsibility between the city government, businesses, and residents.

One example of this is the "Neighborhood Watch 2.0" initiative, where cities empower residents with tools and platforms to report suspicious activities and collaborate with law enforcement. This program goes beyond traditional neighborhood watch programs by incorporating digital platforms that allow real-time communication and data sharing between residents and police.

Additionally, the concept of co-production, where citizens are active participants in the creation of public safety solutions, is gaining traction. This approach encourages residents to work alongside public safety officials in identifying local safety issues and developing tailored solutions. For instance, some cities have established citizen advisory boards that provide input on policing strategies and community relations, ensuring that public safety efforts align with community needs and values.

Public-Private Partnerships for Enhanced Safety

Entrepreneurial cities leverage public-private partnerships to enhance public safety. These partnerships can provide cities with access to cutting-edge technologies and expertise that would otherwise be unaffordable or unavailable.

For example, many cities have partnered with private security companies to expand their surveillance networks. These partnerships allow cities to extend their surveillance capabilities without the need for significant public investment. The private sector can also contribute to public safety through innovations such as smart street lighting, which can adjust its brightness based on foot traffic, reducing energy costs while enhancing safety in public spaces.

In another instance, some cities have collaborated with technology companies to develop apps that connect residents with emergency services more efficiently. These apps allow users to report incidents, share their location with first responders, and receive real-time updates during emergencies.

Building Resilience Through Innovation

An entrepreneurial approach to public safety also emphasizes resilience—ensuring that cities can not only respond to emergencies but also recover and adapt to future challenges. Resilient cities invest in smart infrastructure, disaster preparedness, and continuous improvement of their public safety systems.

For example, cities prone to natural disasters, such as San Francisco, have implemented comprehensive disaster response plans that integrate technology, community engagement, and private sector resources. These plans include automated early warning systems, real-time disaster monitoring, and robust communication networks that ensure residents receive timely and accurate information.

Conclusion

The evolving landscape of public safety demands more than just traditional methods; it calls for an entrepreneurial mindset that embraces innovation, collaboration, and proactive engagement. As cities face increasingly complex challenges, from technological advancements to societal shifts, the need to rethink public safety strategies becomes paramount. By integrating cutting-edge technologies like drones, cameras, sensors, and AI, cities can enhance their ability to predict, prevent, and respond to emergencies in ways that were once unimaginable. Moreover, fostering strong community partnerships and leveraging the expertise of the private sector can amplify these efforts, creating a more resilient and adaptable public safety framework.

In the entrepreneurial city, public safety is not a static service but a dynamic, evolving process that involves every stakeholder—from government officials and law enforcement to businesses and residents. By championing these forward-thinking strategies, we can transform public safety from a reactive function into a proactive force that not only responds to crises but anticipates them, ensuring that our communities are safer, smarter, and more resilient for generations to come. As leaders, we must drive this transformation, paving the way for a future where innovation and collaboration are at the heart of public safety.

CHAPTER 10
DATA-DRIVEN CITIES

"We are surrounded by data but starved for insight."
Jay Baer, marketing expert

The quote "We are surrounded by data but starved for insight" by Jay Baer is an apt starting point for this chapter, reflecting a challenge many cities face today. With the sheer volume of data available, from public safety statistics to real-time transit data and even public sentiment analysis from social media, the potential to transform city management is immense. However, cities often lack the tools or the strategic vision to convert this data into actionable insights.

Reflecting on my own experience, the introduction of Google Street View was a revelation for me as a city leader. It provided a virtual lens into every corner of my city, offering a perspective that was both broad in scope and detailed in observation. This tool allowed us to monitor problematic properties and gauge the effectiveness of our interventions. We could also analyze development patterns to understand better where growth was happening and how it affected infrastructure needs, such as street maintenance and sidewalk repairs.

In my 27 years as a mayor and council member, I frequently pondered how to leverage data to enhance decision-making processes. For instance, using accident data from the police department to identify and address high-risk areas was a recurring thought. By layering this data with information on traffic flows, lighting conditions, and road quality, we could develop targeted strategies to reduce accidents. Similarly, mapping the locations of fires and their frequency helped us identify potential problem areas, whether due to building code violations, high-density housing, or other factors. This information could then inform fire safety initiatives and zoning regulations.

Teachable Moment:

One of the most valuable lessons I learned about the importance of committing to data-driven strategies came when we invested in digital parking meters. The sales representative emphasized the software's ability to provide detailed data on revenue, peak usage times, and meter performance—information that could have helped us optimize meter placement and significantly increase city revenue. Despite the software's $700 annual cost, I was eager about its potential to revolutionize our parking operations and make more informed decisions. However, a few months after the installation, I was disappointed to find that the staff chose not to invest in the software, using the funds instead to purchase an additional meter. This decision revealed a common pitfall in city government: the preference for short-term, tangible gains over strategic investments in data solutions that could provide substantial long-term benefits. In hindsight, this experience taught me the critical importance of aligning city leadership with a commitment to data-driven decisions to avoid missing out on the transformative potential of these technologies.

The Need for a Strategic Data-Driven Approach

To truly become data-driven, cities must develop a strategic approach that goes beyond just collecting information. This approach should involve:

1. **Identifying Key Metrics**: Cities must determine what data is most relevant to their goals. Whether it's tracking crime rates, monitoring traffic patterns, or assessing air quality, focusing on key metrics can provide clearer insights.

2. **Investing in Technology and Training**: Acquiring the necessary tools and training staff to use them effectively is crucial. As seen in the parking meter example, without proper investment in technology and a clear understanding of its value, cities may miss out on significant opportunities to optimize their operations.

3. **Creating a Culture of Data Utilization**: Encouraging departments to share data and insights promotes a culture where data-driven decision-making is the norm. Cross-departmental collaboration can uncover new opportunities to use data in innovative ways.

4. **Partnerships and Collaborations**: Cities can also benefit from partnering with universities, private companies, and nonprofits to gain access to additional expertise and data. These partnerships can offer new perspectives and resources that might not be available internally.

5. **Transparency and Engagement**: Data should not only be used internally but also shared with the public to increase transparency and build trust. Engaging citizens through open data initiatives can also crowdsource solutions and provide new insights into community needs and priorities.

By embracing a comprehensive, strategic approach to data, cities can transform their operations and provide better services to their residents. The journey to becoming a data-driven city is not just about technology; it's about changing mindsets, investing in the future, and committing to a culture of continuous improvement.

Leveraging Expertise:
A Case Study with HdL Companies

Introduction

One example of a company helping cities harness the power of data is HdL Companies, based in Brea, California. Before HdL Companies acquired DataMax at the beginning of 2024, I worked as an advisor to DataMax, a North Carolina-based company, assisting with revenue enhancement services to cities in West Virginia, South Carolina, and Kentucky. The acquisition has been exciting to witness because of the additional services and capabilities HdL Companies brings to the table. HdL specializes in data analysis, providing cities with economic projections, revenue enhancement strategies, and insights to assist in economic development. By collecting and analyzing data on sales tax, property tax, business licenses, and other municipal revenues, HdL helps cities identify trends and opportunities for growth.

HdL's Data-Driven Approach

HdL's approach demonstrates the value of using specialized expertise to navigate complex data landscapes. For example, they assist cities in understanding which business sectors are growing or declining, allowing city planners to adjust economic strategies accordingly. This type of data-driven insight can also help local governments allocate resources more effectively, develop targeted incentives, and ensure compliance with tax laws. All of these are crucial for maintaining and expanding the city's economic base.

Moreover, HdL's data analytics support cities in projecting future revenue streams more accurately, enabling more strategic budget planning and financial stability. This is particularly important in times of economic uncertainty, where having a clear understanding of potential revenue fluctuations can help cities make more informed decisions about spending and saving.

The Benefits of Partnership

By partnering with companies like HdL, cities can leverage advanced data analytics tools and methodologies without needing to build these capabilities in-house. This approach not only saves time and resources but also allows city officials to focus on interpreting and acting on the insights provided.

Conclusion

The experience with HdL Companies underscores the importance of collaboration between local governments and private sector partners to enhance data-driven governance. By combining internal data collection efforts with external expertise, cities can unlock the full potential of their data, driving smarter policies and more effective governance. This collaborative model is a cornerstone of the entrepreneurial approach to city management, where innovation and strategic partnerships pave the way for more dynamic and responsive local governments.

CHAPTER 11

CITY WEBSITES AS ENGINES OF ENTREPRENEURIAL ENGAGEMENT

> *"No one comes to your website to be entertained. They have questions they think you can answer. Content answers questions."*
> **Jay Baer**

Imagine going on a trip and logging into the airline website and being unable to find the tab that says, "Book a Flight" or logging into a hotel website and not seeing the "Book a Room" tab on their website. I know that this seems absurd, but this is a common issue with city websites. A city website is often the first impression that many residents, businesses, and tourists get of a city, making it as crucial as a booking system for airlines and hotels. Just as a traveler would abandon a confusing airline website, residents and visitors may become frustrated and disengaged when they can't easily navigate a city's website. As someone who visits hundreds of city websites each year, it is often impossible to find basic information and often requires digging through several pages just to find out when the council meetings are taking place or to get an address or phone number for a city staff member. On many city websites, the elected officials are listed among the departments, and it is sometimes impossible to find out who is the mayor or city manager. I joke that the person who runs the website is called the "Webmaster", so being the mayor or city manager is just not that important.

Businesses have discovered that just having a website is not enough. It needs to be professionally created and closely managed in collaboration with a responsible individual within the city. Too often, the website's responsibility is a low priority and seldom visited by staff or elected officials. In today's digital age, a city government's website serves as a vital communication and service delivery tool. It is often the first point of contact for residents, businesses, and visitors seeking information or services. A well-designed and maintained website can enhance transparency, improve efficiency, and foster greater civic engagement. Conversely, a poorly managed site can lead to frustration, mistrust, and inefficiency.

Many years ago, before city websites were even a concept, I came across a fellow from another county who served on the local school board. He had a website dedicated to sharing school board news. It was simple, mostly text-based, and not particularly attractive, but it left a lasting impression on me. I later met him at a meeting, introduced myself, and mentioned that I regularly visited his website. He was pleasantly surprised and proud to tell me that he was one of the first local officials in the state to have a website. His name was David Kurtz, and to this day, we remain friends on Facebook. Inspired by David, I developed my first website to communicate city news, years before the City of Clarksburg launched its website.

"That modest beginning made me realize the importance of having a city website as a vital investment for the community. Keeping it up to date became crucial for ensuring citizens could rely on the information provided. Many times, people approached me to express their gratitude for keeping them informed about events or meetings in the community. It marked a turning point for the city, as we embraced the potential of digital communication."

A CASE STUDY OF ELKINS, WEST VIRGINIA

An example of a well-designed website that accomplishes the goals of the city can be found in the City of Elkins, West Virginia. (www. cityofelkinswv.com) This small town in the hills of West Virginia is undergoing a revival and their website is part of the reason. Elkins had prospered for many years by being the center of the lumber industry in West Virginia and the economy was healthy and growing. For years, the Elkin's Forest Festival was a popular event, drawing tens of thousands of visitors each year. As the lumber industry waned, the city began to look a little ragged, and many downtown businesses either closed or relocated to locations along a newly built highway. While they were located close to some ski slopes, most of the traffic stayed on the highway and stopped at the fast food and gas stations along the highway. Some entrepreneurs saw the city as an undiscovered gem and began to open brewpubs and craft beer breweries, along with antique shops and art galleries. The city leaders started to clean up the downtown and invest in upgraded infrastructure to support the new businesses. An empty hotel was revitalized, and a new spirit seemed to be on the rise. The city invested in a communication professional to help brand the city and improve its online presence.

The well-designed website is a billboard on the internet for visitors, residents, and prospective businesses and visitors, whose first taste of the city is via the pictures and visitor resources listed on the website.

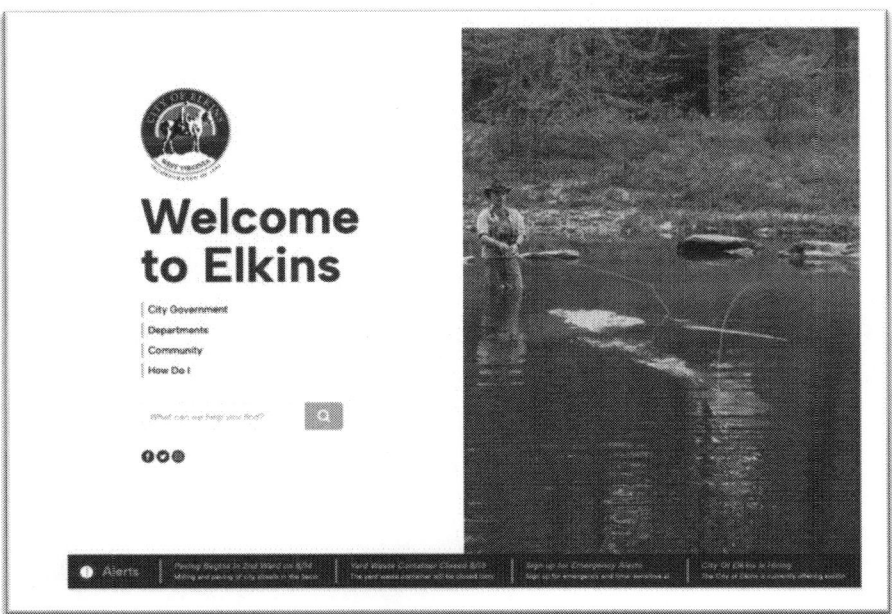

Figure 8 Elkins, WV landing page: A clean, intuitive interface with rotating images and a prominent search bar.

My only suggestion for this landing page is to add the state. Saying Elkins, WV lets people know that they have reached the right page. I often search for a city's website and realize, after some perusing, that I have reached Athens, Georgia rather than Athens, Ohio, or one of the more than twenty "Athens", throughout the United States.

As you move down the Elkins, WV landing page, you find that they continue to add information without looking cluttered or disorganized. They also continue to give helpful navigation to help the visitor access information.

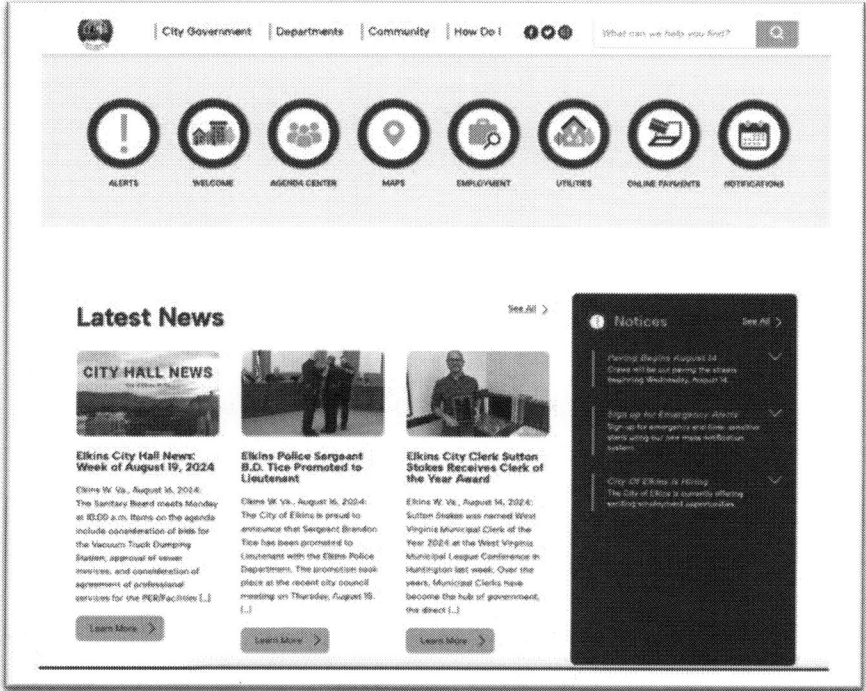

Figure 9 Elkins, WV mid-page layout: Effective navigation with current news and community notices.

When people find the information, they are looking for on a city's website, they are more likely to return to the site when they need information from the city. This can reduce phone calls for routine information and free up city staff. Over my career, I have spoken to dozens of city employees who dread the repetitive calls for simple information, such as office hours at city hall, paving schedules, festival dates, and numerous other inquiries.

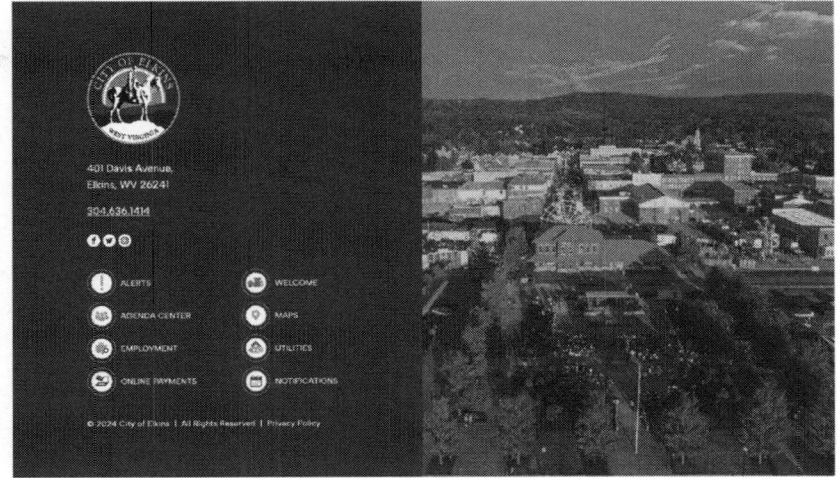

Figure 10 Elkins, WV footer: Essential contact information and social media links, with a 'last modified' date for currency.

One helpful item on the above section of the City of Elkin's landing page is the *"last modified on (date)"*. This gives the visitor confidence that the information is current. I have visited many city websites looking for information, only to discover that they have not updated the information for several months or years. You can bet that visitors will not be relying on that website for reliable information.

I would encourage you to visit the City of Elkin's website and compare it with the far better-known cities of Breckenridge, Colorado, (www.townofbreckenridge.com) and Aspen, Colorado, (www.aspen. gov) and see how an entrepreneurial mindset can compete with some of the best cities and towns in the country.

There are hundreds of examples of well-designed websites to review and compare. The important thing is dedicating time and resources to the design and functionality of your city's website. When I speak to various meetings of the National League of Cities and State Municipal Leagues, I will ask a roomful of elected officials if they have visited their city's website in the last three months and rarely do more than a few hands go up. An entrepreneurial city website not only serves residents better but also positions the city competitively in attracting businesses and tourism.

Some real-life examples of cities that have upgraded their websites and saw measurable improvement in engagement, tourism, or economic development are:

Grand Rapids, Michigan (GrandRapidsMI.gov): The City of Grand Rapids, Michigan partnered with OpenCities, now a part of Granicus to redesign its website and they are seeing improvements in services throughout the city. According to the Granicus website, over 40% of all 311 interactions were "start water and refuse service", that have now transitioned from calls to almost 1000 new online submissions. Also, in just 11 weeks since the new website's launch, the total number of walk-ins to the customer service desk to start water service, dropped by 79%.

Louisville, Kentucky (LouisvilleKy.gov): Louisville redesigned its website with a focus on accessibility, user experience, and improved digital services. They have seen a 25% increase in online service use and a 15% increase in overall site traffic. As a metro government (Louisville-Jefferson County Metro Government), the number and complexity of departments can be a daunting task to consolidate into a single website. In 2022, Louisville won 1st Place in the Overall City Government Experience category in the 6th Annual Government Experience Awards by the Center for Digital Government (CDG).

Waxhaw, North Carolina (www.waxhaw.com): In one of my first podcast episodes for the Amazing Cities and Towns Podcast, I interviewed Jeni Bukolt, founder of **HAVEN** *Creative*® in Charlotte, North Carolina. Jeni is a branding expert and assisted Waxhaw, North Carolina in a complete redesign of their website, as well as the overall brand experience of the city. The city turned around a negative image and went from 108 media inquiries in Fall of 2017, all negative in 2017 to 10 media inquiries, all positive in 2018.

Based on these successful examples, here are the key elements to consider when building an entrepreneurial city website.

1. **User-Friendly Design**
    *   Logical Navigation: Design easy-to-navigate links and put commonly used information on the home page or a prominent tab.
    *   Ensure compatibility: Be sure the site is usable on all devices, including smartphones and tablets.

- Accessibility: Ensure that the site is compliant with the Americans with Disabilities Act (ADA).

2. **Brand Consistency**

- Make sure that the city's website is consistent with the established brand of the city. The use of colors, seals, and styles should follow through all city communication outlets.

- Avoid other city departments creating "off-brand" websites. If the Police, Fire, or Park Departments want to have their "own" website, establish strict guidelines, in order not to degrade the brand integrity.

- Consistency across all city departments strengthens the city's brand and ensures a unified message to residents and visitors.

3. **Timely and Inclusive Content**

- Keep the content current and regularly survey the entire site to make sure unused features or underuse are reviewed and updated as needed.

- Include a "How do I" tab that answers common inquiries.

- Include options for multiple languages.

4. **Social Media Integration**

- Include links to official social media accounts and regularly promote usage.

- Keep track of views, followers, and comments. Have someone responsible for providing timely answers to requests or comments.

- Keep abreast of new social media platforms and review yearly.

5. **Engagement and Transparency**

- Continuously monitor the site to see that each department is providing information to the person who is responsible for the website.

- Protect user data and have clear policies on how user data is collected and protected.

- Have a statement and/or forms for Freedom of Information Act (FOIA) requests.

In today's digital world, being connected to your citizens is essential, and having a user-friendly website can put your city in a positive light and drive tourism and economic development to the city. Improvements in technology have led to some cutting-edge innovations in website creation and management. Integrated AI chatbots, virtual tours, or real-time data dashboards can enhance the user experience.

It is not necessary to have every bell and whistle available, but it is necessary to make the website reflective of the image you want to portray to the public. In some small cities and towns, a part-time webmaster might be a good option but be careful putting the image of your city in the hands of someone who may not share your mission and vision. Keeping information current and relevant is imperative for all-size cities. A communication team should regularly audit the website and make sure that it is achieving the goals set out by the city manager or other administrative professional. Involve the community in aiding the communication team by creating a "Digital Communication Committee" to meet quarterly, to review the current website and social media content and solicit suggestions. In today's remote work environment, many talented and interested people might have valuable insight to make communications better.

A city website is not a one-time thing. It requires updating, innovating, reevaluating, and constant monitoring. Investing in a well-designed website is not just a technical upgrade - it's a strategic move that can position your city for success in the digital age.

CHAPTER 12

ENTREPRENEURIAL REVITALIZATION: THE SUNNYSIDE UP STORY

> *"The creative is the place where no one else has ever been. You have to leave the city of your comfort and go into the wilderness of your intuition. What you'll discover will be wonderful. What you'll discover is yourself."*
> — **Alan Alda**

Sometimes in life, you are at the right place, at the right time, and ready to take on a job that has a 99% chance of failing but is so enticing, that you jump in with both feet. After an exciting adventure of being Mayor of my hometown and becoming President of the National League of Cities, I received a call from Bill Byrne, the Mayor of Morgantown, West Virginia asking if he could meet with me to discuss a position that had just come open in Morgantown. I knew Bill through the West Virginia Municipal League and was intrigued by his description of the job. He explained that it was the Executive Director of the Sunnyside Up-Campus Neighborhoods Revitalization Corporation (CNRC), which was a project to revitalize a student neighborhood in Morgantown, West Virginia. The Board of Directors had not renewed the former executive director's contract and was looking for someone who could shake things up and bring about change in the university city. The Sunnyside neighborhood was once a thriving home to workers in the glass industry who were employed at the Seneca Glass Company, located along Beechurst Avenue in Morgantown, West Virginia. As the factory declined and ultimately closed, the neighborhood undertook a transition from young, working families to older residents trying to maintain large two-story homes. Because of its location close to the campus of West Virginia University, the houses were attractive to property developers who would purchase these large single-family homes and turn them into student rentals. Some longtime residents of Sunnyside were often forced to move due to the difficulty of living next to rambunctious college students. Grant Street, which ran through the center of the neighborhood, became a place for raucous parties and street fires that threatened the safety of residents, living in these crowded wooden structures. Police intervention was a frequent occurrence, and the lawless behavior was impacting the image and safety of the university.

A group of leaders from the City of Morgantown and West Virginia University visited the campus of Ohio State University in Columbus, Ohio around 2004 to see an effort there to address neighborhood revitalization and student housing. Impressed with what they saw, a committee was formed, and the City and University committed funding to do a strategic plan and implement the plan through an appointed board that was not controlled by either the city or the university. An executive director was hired, and offices were rented in the building that previously was the Seneca Glass Company, complete with the old furnaces.

The strategic plan detailed the steps needed to revitalize the neighborhood and was described to me as a working document that would be reevaluated yearly to adjust to any changing conditions. I was excited to have such a well-thought-out document. In my experience, strategic plans would often find themselves gathering dust on a shelf. The Sunnyside Up Strategic Plan focused on improving the existing housing and developing new campus housing and amenities to enhance the neighborhood. The plan was not as strong on engaging the residents, but our board supported my initial efforts of engaging the, mostly student, residents and it was incorporated in the update after the first year.

Sunnyside was home to several bars during my time at WVU, but nothing near the uncontrolled chaos now. A tradition in the Sunnyside neighborhood in the old days was to burn a couch after a big football victory and it always made the student newspaper when it happened. Now, the smoke would barely clear from one fire before another was being set.

Figure 11 One of over 100 burnt dumpsters in the Sunnyside neighborhood.

I drove up to Morgantown to visit the neighborhood that I had not seen since my college days and my first impression was that it looked like a war zone and was clearly in crisis. Of the over 100 trash dumpsters behind the houses, I only counted two that had not been burned and graffitied. The houses were a mix of well-maintained single-family homes and dilapidated student housing with broken sidewalks and multiple cars parked wherever there was an opening. As I drove down the streets, I often had to wait for groups of students who were throwing a football or just hanging around drinking out of red Solo cups. I had been an elected official (mayor and councilmember) in a city about 30 miles south of Morgantown for about twenty years at the time and was also working for the West Virginia Housing Development Fund, a quasi-governmental agency that offered mortgage bonds and did revitalization work throughout the state. Although I was an alumnus of WVU, I was thinking about passing on this position.

The Board of Directors was a great group of people, composed of landlords, college administrators, city officials, and business leaders and they said that they were willing to be innovative in approaching the challenge. I insisted on a three-year contract and their promise to give me leeway as we moved forward. I laid out an entrepreneurial approach to addressing the issues in the neighborhood and I told them that I would need the residents (mostly college students) of the neighborhood to be the building blocks that we needed to get the property owners and developers on board. I can remember the quizzical looks on their faces, thinking that these undisciplined college students were going to be the leaders of the revitalization.

Roof collapses on Grant Ave., leaves two injured

By Kellen Henry
Staff Writer

Two people were transported to Ruby Memorial Hospital after a roof collapsed on Grant Avenue Saturday evening.

About seven people were on the porch roof of 426 Grant Ave. when it fell after 8 p.m. said Captain Mark Paravasos of the Morgantown Fire Department.

The roof collapsed because of major structural problems with the house, including shifting in the walls and basement.

"I'm sure the weight of the people on the roof didn't help. It probably added significantly to the weight the roof was holding," Paravasos said.

The fire department and city building code division condemned that house along with 430 Grant Ave., 434 Grant Ave., and 438 Grant Ave. because of structural problems

noticed while at the scene.

Neighbors said they saw people standing and walking on the porch shortly before it fell from the second story.

"There was lots of smoke and dust. I thought the whole house collapsed," said junior John Lopacki, a Grant Avenue resident. "It was pretty loud."

The fire department, EMTs and police responded within 10 minutes, he said.

Neighbor Tim Hiera, a senior, said he saw emergency workers treating two injured people after the collapse.

The names and conditions of the two transported to the hospital have not been released.

The Grant Avenue house is owned by Bill Burton of Bel-Cross Properties.

danewsroom@mail.wvu.edu

Taped off and condemned, 426 Grant Ave. sent two to the hospital Saturday night after the roof porch collapsed. The house was owned by Bill Burton of Bel Cross Properties.

KENDAL MONTGOMERY/THE DAILY ATHENAEUM

While the Sunnyside Up Board of Directors was supportive, they were receiving a good deal of input from a variety of interest groups. There was a feeling that investing in a dilapidated neighborhood was not worth the effort and the money could be better used elsewhere. Like many college communities, friction between the "town-gown" residents was a daily occurrence and students were sometimes thought of as an unnecessary inconvenience. I looked for help and we found the International Town-Gown Association and became active members and were able to meet with other college communities and share success stories and challenges. This organization is a great resource, and the networking opportunities were a valuable part of the membership.

As in many success stories, luck is a great partner. Before I had officially started work, I was getting my office in shape, and I heard a knock on the door. A young lady asked if this was the office of Sunnyside Up.

She explained that she was a member of the Student Government, and her platform was addressing the housing issues throughout the campus. Her name was Erica Rogers, and I told her I had not officially started the job but was willing to talk with her. Her youthful energy came through with every sentence and she told me that she could introduce me to the Student Body President and invited me to the first meeting of the student government. I drove home that night thinking that this just might be possible.

I knew that the first thing we needed to start on was the deplorable condition of the neighborhood. Graffiti and mounds of garbage were on every corner and weeds and overgrown trees made the streets look foreboding. I attended the first meeting of the student government, and they were excited to work with me in cleaning up the notorious neighborhood. The student body president, an impressive young man named Jason Parsons, assigned one of his assistants to work with me. It was so much fun to be the head of this youthful start-up organization, and I had never had such enthusiasm during my previous experiences in local government.

The assistant was a MPA graduate student named Nelson France, who was interested in local government. When he saw the pictures of my local government career and the people and places I had been, he became a regular visitor to my office for long discussions of government, politics, and life. He volunteered to head up a neighborhood cleanup before one of the big football games. Facebook had just gotten popular and although I had an account, I was doing little more than posting family pics or meals that I was eating. Nelson said that he would organize a Facebook event, and students would sign up. I was a little reluctant, but he assured me that this would work. I called him on the Monday before the Saturday clean-up and asked him how many people had signed up. Thinking I would need a dozen or so pairs of gloves and a couple of boxes of trash bags, I was stunned when he said that he had 320 sign-ups and was expecting several more before Saturday. I rushed to the Lowes Home Improvement store and bought every pair of work gloves they had and two cases of heavy black garbage bags.

As Saturday approached, Nelson came to my office and asked how we were going to organize the event, and we came up with a plan to appoint the student government members as team leaders and divide the students into groups of twenty. I drove to the neighborhood on Saturday morning at about 7:00 AM and Nelson and a few of the student government kids were milling around. Disappointed, I asked Nelson where everyone was. He let me know that college students don't jump out of bed till it's time to go and that he was sure they were coming. When the clock hit 8:00 AM, it looked like the movie, Children of the Corn, as sleepy students were emerging from every direction. My fears were quelled!

Figure 12 WVU students helping clean up the Sunnyside neighborhood.

I specifically remember an elderly lady who had lived in her well-kept house for over sixty years. She saw a group of students during one of our clean-ups and invited them up to her porch for iced tea. She was so excited to help and told us how she and her husband had built the house and raised their family. She was very careful with her words and did not want to offend the students sipping iced tea, but she expressed remorse at the decline in the neighborhood. It was a touching moment when the students offered to pick up around her property and said that they would keep an eye out for her and stop by to visit when they could.

The Sunnyside Clean-Up made the front page of the papers, and the television stations interviewed the student volunteers and put them on the evening news. Social media also caught fire and pictures of the students hauling old air conditioners and broken lawn chairs were shared throughout the campus. We also had a lot of students asking to take pictures with me, in front of the Sunnyside sign to let their parents know that they were doing something good. After we were done, the streets and right-a-ways were clean and there were mounds of black garbage bags waiting for the city public works department to pick them up on Monday morning. When I called the city manager to let them know, he was astounded by how many bags of debris had been picked up.

My Board of Directors was pleased with the positive publicity but were still anxious to do something about the burnt dumpsters and delipidated houses. With Parent's weekend approaching, I asked the waste hauler if we could put a burned-out dumpster on a parking lot and paint it, as something the parents could do with their students, and they agreed. I bought several gallons of blue and gold paint, which are the WVU colors and a handful of rollers and paintbrushes. I had planned on this taking about an hour on my Saturday morning and I told my wife I would be home early to watch some college football on TV. As the parents assembled around 9:00 AM, they were dressed in old clothes and ready to paint. The group of about twenty-five parents and students had completed the first dumpster and were asking if they could go up the alleyways and paint other dumpsters. I had not discussed this with the waste company, but I figured it would not upset them for the parents and students to paint their dumpsters. By 11:30 AM, the parents and students had painted twenty-eight dumpsters and only quit because we had run out of paint. Once again, the Sunnyside project was front-page news, and we were on our way to improving the neighborhood.

I had read about some street artists in Europe doing guerilla projects at night, such as cleaning up an abandoned lot and the residents awakening to a new park with flowers and benches made from old tires or pallets. I knew this would appeal to these kids and we started painting dumpsters at night and putting a stencil of "Sunnyside Up" on each one in bright gold paint. Soon, over 100 dumpsters were completely painted and became a popular item in the neighborhood. We also hired a tree company to cut the overgrown right-a-ways, while the students installed flowers and painted rocks along the border. Once the residents saw a clean, mowed green space outside their apartments, they moved out of the streets and the "pocket parks" became a social hangout.

The guerilla-style projects quickly gained traction, attracting volunteers from local Boy Scout troops and church groups who were eager to join in our unconventional approach to community improvement. These efforts had unexpected benefits, as they fostered connections between the young college students and parents with children in the community—two groups that had previously been somewhat disconnected. Many community members had steered clear of the student-dominated areas due to concerns about behavior, but these collaborative projects helped to break down those barriers. The shared work created a sense of camaraderie and mutual respect, bringing the broader community together in a way that had not been seen in years.

Figure 13 Student volunteers painting burnt dumpsters and stenciling the Sunnyside logo

108

Another resource that was instrumental to our success was the WVU Landscape Architecture Department. Ashley Kyber, a landscape architecture professor had assigned her students a semester-long project to develop unique outdoor art and one of the students showed up at my door to see if we would work with him on a project. Soon we met with Professor Kyber and offered up the Sunnyside neighborhood for a competition among the class. I attended several classroom discussions about the projects, and we identified sites in the neighborhood where they could be displayed at the end of the semester. While most of the projects were temporary in nature, it was an opportunity to bring attention to the neighborhood and engage the students in a fun and exciting project. The public art projects stayed up for several weeks and the Sunnyside neighborhood received a lot of attention and news coverage. An additional benefit of working with the landscape architecture department was that we learned that WVU had some of the most sophisticated computer drafting tools and we were able to save significant dollars preparing professional plans and schematics for upcoming student-led projects.

Graffiti was a significant problem in the neighborhood, and we grappled with it constantly. In addition to the trash dumpsters, almost anything with a flat surface had been spraypainted and defaced. Fixing this problem was going to take some innovation and hard work. We developed a system where we would have paint mixed to match the cement color through a new color-match technique at the home improvement store. Being a stickler for detail, I would insist that our student volunteers paint the entire wall and not leave evidence that it had been tagged. While this did not eliminate the problem, when property owners saw that we were taking an interest, they tended to install outdoor lighting and monitor their properties more closely. On a side note, the paint matching was so good that most people had no idea the cement walls had even been painted.

All this work was done without involving the code department and no citations were being issued for grass and weed violations, which pleased the property owners. One property owner, who had not been a supporter of Sunnyside Up, offered up his dump truck and told me we could use it anytime we needed. Yes, there were still an occasional dumpster fire and litter was a constant problem, but it was manageable. After each football weekend, I would drive through the neighborhood and not any graffiti or burnt dumpster and we dispatched a temp worker that the board allowed me to hire.

WVU students pledge to volunteer

Will do $100K worth of community service during the next year

INFO: WVU Student Government Association, sga.wvu.edu; Sunnyside Up, sunnysideup wv.org.

BY CASSIE SHANER
The Dominion Post

The WVU Student Government Association announced plans Wednesday to perform volunteer service worth $100,000 during the next year — in an effort to give back to the Morgantown community and strengthen town-gown ties.

SGA President Jason Parsons and Sunnyside Up Executive Director Jim Hunt signed a memorandum of understanding Wednesday in which the two groups agreed to work together to improve Sunnyside. The document, part of a new SGA community initiative, was signed near Summit Hall, where construction continued on WVU's new honors dorm — another joint project involving the two groups.

Parsons also pledged to work with Sunnyside Up and the WVU Center for Civic Engagement to help students complete the $100,000 in community service in Sunnyside and other areas through next September.

The initiative is designed to encourage students to fulfill their responsibilities to the communi-

it," Colebank said. "We'll fill in the gaps.

Morgantown Mayor Ron Justice said students have already committed to perform 900 hours of community service through the Center for Civic Engagement.

He also noted that the announcement of SGA's community initiative is timely, as Morgantown City Council agreed Tuesday to send a tax-increment financing plan for Sunnyside to the West Virginia Development Office. Justice said the city is committed to making the changes in Sunnyside a success.

Bob Gay/The Dominion Post photos photos

WVU senior forestry major Will Straub (above) does his part to promote positive feelings in his Sunnyside neighborhood through this large banner on the front of his Grant Avenue apartment. WVU's

Figure 14 WVU Students step up to volunteer for community service.

The student government participation grew with each school year and over one hundred student government students participated in the project over the five years I was executive director. Erica Rogers went on to law school and is now Policy Director at the City and County of Denver, Colorado. Student body President Jason Parsons went on to become President and CEO of a hospice agency in Virginia. Nelson France is now a community relations director for a Jewish organization in Northeast Florida. The student body President who followed Jason Parson and was instrumental in keeping student government engaged in the project was a young man named Chris Lewellen and we have become great friends. Dozens more of the now-graduates are spread out across the country doing great work and keeping in touch with me through Facebook and LinkedIn. To have the opportunity to work with these talented students was an honor and privilege for me and I will be eternally grateful to have met each one of them.

While much of this work was fixing and improving existing buildings and infrastructure, we also focused on the larger infrastructure projects by forming the first Non-Profit Tax Increment Financing (TIF) in the state. Tax Increment Financing is a law that allows developers to utilize the "increment" between the existing property value and the improved value after improvements have been made. It creates a value that can be monetized through a bond issue and allows the developer to invest in new infrastructure, paid for by the bond proceeds. We installed new sidewalks; LED streetlamps and decorative bus stop shelters which were designed to survive the inevitable student abuse. The use of the TIF is usually done by a for-profit developer and through some study and assistance from a talented legal team, we utilized this innovative approach to invest in the neighborhood.

Another project incentivized property owners to have architectural plans for improvements and renovations. Our board of directors approved funding for a matching grant program that paid for the plans and gave owners valuable suggestions to increase parking and get more functionality on their projects. While even our board was skeptical, this proved popular with the property owners and small developers, and we regularly exceeded the demand for these matching grants.

The Sunnyside Up experience stands out as one of the most rewarding chapters of my career. It provided me with a living laboratory to explore and apply entrepreneurial thinking within the local government sector. Few people ever have the chance to bring their ideas to life in such an inspiring and impactful project. The lessons I learned from this experience have continued to guide me as I work with communities across the country. Stepping outside conventional thinking and adopting an entrepreneurial approach to familiar challenges faced by cities and towns is not without risk. There were many evenings when I lost sleep, concerned that my 'innovative' approaches might backfire or even lead to my dismissal. However, those risks were integral to the journey, and they ultimately played a crucial role in the project's success.

I would be remiss if I did not acknowledge the invaluable contributions of our dedicated board of directors, the financial support from the city and university, and the incredible young people who embraced a vision for a better community. It is heartening to know that many of our student government leaders, interns, volunteers, and others have gone on to serve in public service, continuing to champion new and innovative ways of addressing local government issues.

CHAPTER 13

EMPOWERING CITIES: THE BEARING ADVISORS MODEL FOR INNOVATION

> *"At Bearing Advisors, we don't just consult—we collaborate, innovate, and empower. Our mission is to transform cities into dynamic hubs of opportunity by harnessing entrepreneurial thinking and cutting-edge strategies."*
> **Phil Riley, Co-Founder of Bearing Advisors**

BEARING ADVISORS

As I reflect on my years of service in local government, one thing has always been clear: the most successful communities are those that embrace innovation and entrepreneurial thinking. This realization led me to join forces in 2020 with five like-minded local government professionals to form Bearing Advisors, LLC. We recognized a growing need for a consulting firm capable of bridging the gap between traditional governance and the evolving demands of the 21st century. Together, we set out to help cities and corporations navigate the complex landscape of local government with creativity and confidence.

In this chapter, I will share stories from our journey—challenges met with ingenuity, partnerships forged in the spirit of collaboration, and successes that have left a lasting impact on the communities we serve. Through these experiences, I hope to illustrate how Bearing Advisors has become a trusted partner in driving change and fostering entrepreneurial leadership in local governments across the country.

The Bearing Advisors Team:

The Bearing Advisors team is a powerhouse of experience and expertise, uniquely equipped to drive innovation in local government. Comprising leaders with decades of experience across various sectors, each member brings a distinct skill set that complements and enhances the team's collective capabilities. Phil Riley's visionary leadership in public-private partnerships, particularly his groundbreaking work with Utility Service Partners, exemplifies the kind of strategic thinking that underpins Bearing Advisors' approach. Cathy Spain, with her extensive background in government finance and intergovernmental relations, provides the team with unparalleled insights into the fiscal challenges cities face and the innovative solutions needed to overcome them. Brad Carmichael's expertise in business development and his role in creating successful partnership models ensure that Bearing Advisors can forge strong, sustainable relationships between cities and the private sector. Mike Conduff's deep understanding of governance, honed through years of city management and leadership in policy governance, adds a crucial dimension of strategic oversight, ensuring that all initiatives are aligned with the best practices in public administration. Mike Madden's sharp business acumen and his success in guiding organizations through transformational change ensure that Bearing Advisors can help cities navigate complex challenges with agility and confidence. Finally, Jim Hunt's extensive experience in municipal leadership and his commitment to fostering inclusive, innovative communities anchor the team's efforts in practical, impactful outcomes. Together, this diverse team of professionals offers a holistic approach to modernizing local government, combining strategic vision, operational expertise, and a deep commitment to public service.

Navigating the Local Government Culture:

One of the early insights for the Bearing Advisors team was the realization that many companies were still relying on outdated sales strategies. Historically, the relationship between vendors and cities was largely transactional. Cities would notify vendors when they needed to replenish items like water meters or dump trucks. Vendors, in turn, would review the specifications and bidding requirements, submit their proposals, and wait for the award date. Sales in this context often involved nothing more than periodically checking in with a city's department head or foreman over coffee, gauging what was in the budget or what needed urgent repair.

However, the introduction of new technologies began to shift this landscape. The straightforward commodity purchases of the past were replaced by more complex services and equipment that didn't align with traditional city purchasing processes. This evolution led cities to explore joint purchasing agreements, allowing one city to bid on an item and enabling other cities—regionally, statewide, or even nationally—to procure the same item without undergoing a separate bidding process. As the pace of business accelerated, the typical 90-day bid cycle felt increasingly outdated, especially in a world where businesses could receive deliveries from Amazon and other suppliers in days, or even hours.

The nature of products and services also evolved, with companies offering SaaS (Software as a Service) solutions facing significant challenges in working with cities. Specifications often became so tailored that only one supplier could meet the needs, leading cities to declare nearly everything an 'emergency' purchase. Additionally, the rise of online sales and a more transient city workforce made it harder for sales professionals to build relationships with city personnel. As a result, many found themselves shut out before they even had the chance to make their pitch.

Recognizing these challenges, the Bearing Advisors team identified a critical need for updated training. Companies needed to understand the complexities of city government hierarchies and develop the skills necessary to navigate past gatekeepers effectively when presenting their products and services.

Breaking Down Barriers to Innovation and Technology:

An interesting conundrum often arises between cities and vendors: despite the availability of innovative products and services that could save cities time and resources, these solutions are often inaccessible due to the constraints of the existing purchasing systems. For example, I recall working with an LED streetlight company that offered to replace a city's old, inefficient streetlights and finance the project through the resulting energy savings. Yet, many cities had already entered into long-term agreements with their power suppliers to provide and maintain streetlights at a fixed tariff. As a result, millions of dollars were being spent annually on outdated lights that frequently burned out, while cities missed out on the cost savings and improved efficiency that LED technology could provide, with its 20-year life expectancy and reduced maintenance needs.

The Bearing Team recognized that overcoming these entrenched practices required a strategic shift in approach. We advocated for a top-down sales strategy, emphasizing the importance of building relationships with policymakers and administrative leaders who had the authority to challenge and revise outdated procurement agreements. In the case of the LED streetlights, it was essential to educate elected officials about the restrictive contracts that hindered innovation and to demonstrate the substantial long-term savings that could be realized by adopting the new technology.

Role-Playing and Finding Champions

One of our key strategies to assist both cities and companies was the use of role-playing scenarios designed to help sales professionals effectively engage with elected officials and senior administrators. Leveraging the experience of team members who had served as both Mayors and City Managers, we crafted realistic scenarios that simulated speaking at council meetings or interacting with officials at conferences. A common misconception among many sales professionals was that securing the mayor's approval means the sale is assured. However, in many cities, it is the City Manager who prepares the budget and makes the final purchasing decisions. Understanding this dynamic is crucial for successful negotiations.

Additionally, we introduced the concept of "Champions"-key individuals within the city government who understand the value of a product or service and can advocate for it within the organization. The champion, often a member of the governing body or knowledge-able staff member, play a vital role in educating other decision-makers who may be less familiar with the innovative solutions being proposed. While it's relatively straightforward to explain the benefits of Tasers or protective vests for the police department, more complex innovations, such as using drones as first responders, require a champion who can articulate the capabilities and advantages of the technology in a way that resonates with the broader decision-making body.

Brand Licensing and Innovative Partnerships:

A trend that the Bearing Advisors team is seeing is the emergence of new ways of providing products and services to local government.

Two members of the Bearing Advisors team pioneered an inno-vative brand licensing program by partnering with cities to promote a utility warranty program. The program was a win-win for the cities and the warranty company since it provided needed protection for a utility service line that could have cost the homeowner many thousands of dollars if they failed. In exchange for using the city's logo on their mailings, the company paid a brand license fee, offsetting any cost for the program and providing money for city needs.

Another example is the town of Cooperstown, New York, home to the Baseball Hall of Fame. A small town of only 1,848 residents, Cooperstown has successfully leveraged its connection to baseball history by forming partnerships with sports brands and memorabilia companies. The annual Hall of Fame Induction Weekend draws thou-sands of visitors, providing an economic boost to local businesses. These partnerships not only boost the town's brand but create a steady stream of revenue and economic impact.

By working with cities and towns to find the 'right' partnership or opportunity, Bearing Advisors works to ensure that the city is protect-ed and that all parties adhere to the highest ethical standards. These relationships have brought significant funds into city and town coffers and provided needed revenue for critical local needs.

Working with State and National Associations:

The value of working with State and National Associations is a specialty of the Bearing Advisors team and one that pays dividends for innovative companies and cities and towns. Having Bearing Advisors team members who have served in leadership roles and senior staff positions for both state and national associations brings value and strategic advice.

Organizations like the National League of Cities, the International City/County Management Association, the Government Finance Officers Association, and the 49 State Municipal Leagues assemble tens of thousands of city and town officials and administrative staff on an annual basis and promote networking opportunities for innovative companies looking to introduce themselves to these officials.

By maintaining close professional relationships with the staff of these state and national associations, Bearing Advisors has facilitated partnerships and affiliations that bring needed revenue for programming activities to these associations.

Conclusion:

Through our work at Bearing Advisors, we have seen firsthand how entrepreneurial thinking can transform local governments from within, breaking down outdated practices and building smarter, more responsive cities. By fostering innovation, forming strategic partnerships, and empowering local leaders with the tools and insights they need to navigate complex challenges, we are helping to create a new paradigm for governance—one that thrives on adaptability, collaboration, and a relentless pursuit of opportunity. This entrepreneurial approach is not just a strategy; it is a mindset that can propel any city into a future of sustainable growth and resilience. As cities around the country look to evolve and better serve their communities, Bearing Advisors remains committed to guiding them on this journey, proving that with the right vision and leadership, any city can be an 'entrepreneurial city.'"

CHAPTER 14
UNLOCKING PUBLIC/PRIVATE PARTNERSHIPS IN THE ENTREPRENEURIAL CITY

"By matching private and public sectors in partnership, we can break down barriers to sustainable growth, which no sector can do on its own."
Lars Lokke Rasmussen, Prime Minister of Denmark

Introduction:

Throughout my career in local government, I've encountered many public-private partnerships, each with its own set of lessons. However, one experience stands out as a cautionary tale for city officials considering such arrangements. One day, while driving through town, I found myself behind an old Dodge Truck with a lightbar and markings resembling a police car. What caught my attention was the bold, black lettering on the back tailgate: "Donated by Tony's Used Cars." Surprised, I immediately called our city manager to inquire if this was indeed one of our police vehicles. After some investigation, he confirmed that it was and explained that a local used truck dealer had donated it to the department. I asked if we had any policies governing such donations or if it was appropriate for a police city vehicle to effectively advertise for a private business. The city manager admitted that we did not have any guidelines in place and would need to draft a policy for city council approval. This incident left me questioning whether this was an example of "entrepreneurial" thinking by our police chief or a potentially problematic blurring of lines between public service and private interest.

I also recall the time that Louisville, Kentucky Mayor Jerry Abramson was approached by KFC, which is headquartered in the city, and wanted to donate money to fill potholes and then place a stencil that read, "Refreshed by KFC". The mayor gained some publicity for the city and filled 350 potholes, however, the local PETA chapter offered to double KFC's donation for a chance to place a stencil that said, "KFC tortures Chickens!" Seems that no good deed goes unpunished.

These somewhat humorous examples point out the necessity of making sure the city or town gets value for their partnerships and that possible "potholes" are discovered before signing the documents. The following paragraphs give an idea of what is needed to properly evaluate public/private agreements.

Evaluating the benefits and liabilities:

Evaluating the benefits and liabilities of a proposed public-private partnership (PPP) involves a thorough analysis of several key factors to ensure that the arrangement is advantageous for the public sector while minimizing risks. Here's a structured approach to consider:

1. Define the Objectives and Scope

   - Benefit: Clearly outline what the partnership aims to achieve, such as infrastructure development, improved public services, or economic growth. Ensuring alignment with the public sector's strategic goals is essential.

   - Liability: Misaligned objectives can lead to conflicts, project delays, or public dissatisfaction if the partnership fails to deliver expected outcomes.

2. Financial Analysis

   - Benefit: Evaluate the cost-effectiveness of the partnership by comparing the financial input from both parties, the potential for cost savings, and revenue generation. Assess whether the partnership allows for leveraging private sector capital to reduce the burden on public finances.

- Liability: Consider the risk of financial over-dependence on the private partner, which could lead to financial instability if the partner faces difficulties. Also, ensure that the public sector retains fair value and does not face hidden costs or unfavorable terms.

3. Risk Assessment

- Benefit: Identify how risks are shared between the public and private sectors, such as construction, operational, financial, or market risks. A well-structured PPP should allocate risks to the party best equipped to manage them, thereby enhancing project success.

- Liability: Poorly defined risk-sharing agreements can lead to disputes, increased costs, or project failure. Assess potential liabilities, including legal, environmental, or reputational risks, and ensure they are adequately mitigated through contracts and insurance.

4. Legal and Regulatory Compliance

- Benefit: Confirm that the partnership complies with all relevant laws, regulations, and standards. Compliance reduces the risk of legal challenges and enhances public trust.

- Liability: Non-compliance or regulatory violations can result in legal penalties, project delays, and damage to public trust and reputation.

5. Quality of Service and Performance Standards

- Benefit: Establish clear performance indicators and service quality standards to ensure that the private partner meets the public sector's expectations. High-quality service delivery can enhance public satisfaction and trust in the partnership.

- Liability: Ambiguous or unenforced performance standards can lead to subpar service delivery, public dissatisfaction, and potential financial losses for the public sector.

6. Public Interest and Community Impact

    - Benefit: Evaluate how the partnership will benefit the community, such as through job creation, enhanced services, or improved infrastructure. Consider any potential social or economic benefits.

    - Liability: Assess any negative impacts on the community, such as displacement, inequitable access to services, or environmental harm. Ensure that the partnership includes measures to address and mitigate any adverse effects.

7. Governance and Accountability

    - Benefit: Strong governance structures ensure transparency, accountability, and effective oversight of the partnership, reducing the risk of corruption and mismanagement.

    - Liability: Weak governance or lack of transparency can lead to corruption, mismanagement, and loss of public trust. Ensure that both partners are held accountable for their roles and responsibilities.

8. Flexibility and Adaptability

    - Benefit: Evaluate the partnership's ability to adapt to changing circumstances, such as economic shifts, technological advancements, or evolving public needs. Flexibility can enhance the partnership's resilience and long-term success.

    - Liability: Rigid agreements may not allow for necessary adjustments, leading to inefficiencies, conflicts, or failure to meet public needs over time.

9. Exit Strategy and Long-Term Sustainability

    - Benefit: Develop a clear exit strategy for both parties, outlining conditions under which the partnership can be dissolved or renegotiated. This ensures that the public sector is not locked into an unfavorable arrangement.

    - Liability: Lack of a clear exit strategy can lead to long-term financial or operational burdens for the public sector, particularly if the partnership becomes unviable or contentious.

Public-private partnerships (PPPs) can range from large-scale projects involving billions of dollars to simpler agreements, such as a local church allowing the city to use its parking lot during the week. An entrepreneurial city approaches these proposals with confidence, conducting thorough evaluations to ensure that each agreement aligns with public goals, delivers net benefits, and serves the community's best interests while effectively managing potential liabilities. Rather than being intimidated by the complexity of these partnerships, the entrepreneurial city uses a careful, strategic approach to make well-informed decisions.

Note of Caution: Before entering into any public-private partnership, it is crucial for cities and towns to seek professional guidance. Public-private partnerships can offer significant benefits, but they also come with complex financial, legal, and operational considerations that require careful planning and expertise. Engaging experienced consultants, legal advisors, and financial analysts can help ensure that all aspects of the partnership are thoroughly evaluated and structured in a way that protects the public interest, mitigates risks, and aligns with the community's long-term goals. Remember, a well-informed approach is key to maximizing the potential benefits of a public-private partnership while safeguarding your city's resources and reputation.

CHAPTER 15

THE UNTAPPED POWER OF SOCIAL MEDIA IN THE ENTREPRENEURIAL CITY

"Goodbye gatekeepers! Social media is transforming the way local governments communicate, bypassing the click-bait headlines of traditional media and forging a new path towards dynamic two-way communication with residents.
Sam Toles, Founder of CiviSocial and former Chief Content Officer at Bleacher Report

Introduction

Over the years, I've watched social media merge with local government. I had the unique experience of being around when there was no social media in local government, and then seeing the emergence of social media and then the pushback against it. As I travel around the country and talk to hundreds of cities, I hear the universal complaint from local government officials that they believe social media has ruined local government. This is a common feeling, but I try to dig a little deeper. Social media is not going anywhere; it will be a part of local government and society for the foreseeable future. With broadband access and the low cost of producing content, cities need to study and find ways to make social media a positive asset to local government. Additionally, entrepreneurial cities have recognized that having an interactive link with their citizens and others who may want to move themselves or their businesses to the city is an valuable tool, that will only get more valuable in the future.

Historical Context

One story I like to tell when speaking about social media is comparing it to the advent of the telephone. When the telephone was first introduced, it seemed like a novelty. Businesses didn't use it much because not many people had phones. Similarly, social media started around 1978 with college professors on Bulletin Board Systems (BBS), sharing files, reading news, and sending messages. and its value was hard to describe at the beginning. In 2003, My Space launched and quickly became the most popular social network. It was especially popular among younger users and musicians. LinkedIn was founded as a professional networking site focused on connecting business professionals. Facebook was followed in 2004 by Mark Zuckerberg at Harvard and expanded to other colleges before being introduced to the public in 2006. However, as more people used the internet to access information, social media became a tool for sharing information digitally, much like how the telephone became a social network.

Initial Resistance and Adoption

Local governments were not early adopters of social media. They struggled to find reasons to be on social media, often due to generational and technological challenges. It took time for local governments to see the value in social media, especially with live streaming and recording meetings. Social media began to expose information that was previously controlled by local media, creating a new dynamic in public information dissemination.

The Role of Social Media in Modern Local Government

Social media has spread rapidly and offers free access to a vast amount of social information. Initially used by young people for communication, it quickly became a platform for sharing pictures, achievements, and events. Local governments, initially hesitant, began to see the benefits of social media for transparency and direct communication with residents. Live streaming meetings have made it easier for citizens to stay informed and engaged and the archived files of meetings can be a useful tool for those who cannot attend meetings, either virtually or in person.

Benefits of Social Media for Local Government

One of the benefits of social media is the ability for the city to put its message out there, live stream events, and archive them for future reference. This transparency is valuable for residents and allows the city to share its successes and events without relying on traditional media. For example, Port St. Lucie, Florida, has a dedicated social media director and a working studio to produce content, showcasing the value of social media in local government.

Challenges and Misconceptions

Despite its benefits, social media has its challenges. Local government officials often complain about misinformation and negative feedback. The reality is that social media allows for anonymous posting and comments from outside the city, creating a different dynamic from traditional public meetings. However, this also means that more voices are being heard, and local governments need to adapt to this new reality.

Case Studies

Port St. Lucie, Florida, serves as a successful example of integrating social media into local government. With a dedicated social media director and a robust strategy, the city has won national awards for its social media footprint. Smaller cities can also be effective with part-time social media directors and inexpensive equipment, showing that size and budget are not barriers to successful social media use.

Beaumont, Texas, is another great example of using social media to benefit the community. Their robust social media footprint highlights newsworthy events, alerts, government meetings, and helpful information for residents. Several years ago, during devasting floods, Councilman Audwin Samuel used the Facebook live streaming feature to connect with citizens and relay important information to the public. Councilman Samuel, whom I met years ago at the National League of

Cities, is an innovator, and when I saw the interaction with citizens in a time of crisis, I knew he was on to something.

Measuring Impact and Effectiveness

To gauge the impact of social media, cities need to measure engagement metrics, including Facebook followers, Twitter interactions, and YouTube views. Different cities show varying levels of engagement across platforms, and understanding these patterns helps tailor social media strategies to each city's unique needs.

I've researched social media usage by cities and towns of various sizes. While not exhaustive, this research provides insight into the national trends. The following charts were compiled by visiting city websites and following their social media links. Since social media numbers fluctuate daily, this data represents a snapshot of the second quarter of 2024.

The research also highlights a fundamental flaw in many social media platforms: they were designed for individual users, and cities and towns often struggle to integrate effectively. Some cities rely on a staff member to create a page on their personal account. When the staff member leaves or loses interest, the page may be deleted or become difficult to find. Additionally, many cities have multiple pages for different departments—such as police, parks, and economic development—making it challenging for residents to find information. If a city opts for multiple pages, the chapter on "Branding" can help ensure a consistent look and feel, making it easier for residents to identify official city pages.

As cities embrace social media, many have found that it is necessary to establish or expand their communication departments. This will vary by size and budget, but I believe that entrepreneurial cities will see the value of the investment and allocate resources appropriately. While not every city or town will have a production studio or complex editing facilities, it is possible to produce professional-looking content with a modest investment. A tip for those looking to develop a larger social media footprint would be to look at what your peer cities are doing. A quick check online can give you valuable insight into what neighboring cities and towns are doing or even compare things like college towns, beach towns, etc.

I recently met with Sam Toles, a former Cathedral City, California councilmember who has had an extraordinary career in the digital world, working for some of the largest digital platforms in the country. Sam was a former Chief Content Officer at Bleacher Report, former general manager of Vimeo, and Head of Digital at MGM Studios. Sam's latest venture is the formation of a company named CiviSocial, which helps cities and counties be more effective on social media. Sam works primarily with cities under 250K and has a program that involves city employees acting as content creators, to feed information to the Chief Information Officer or whoever is responsible for the social media platforms. This might be a code official who is comfortable on camera and pointing out unusual things in the course of their day or a police officer that gives safety tips for back-to-school season. Creating short, informative videos with a lighthearted touch can communicate the city's message without seeming heavy-handed. Sam understands how cities work, and his digital background can empower cities to better communicate with their citizens.

On the following pages is an analysis of the various social media outlets that are most used by cities and towns. Please be advised that with the constantly changing nature of social media, the numbers in the charts may not reflect current activity. The charts are a snapshot in time and are used to illustrate the relative popularity of social media platforms.

FACEBOOK USE IN VARIOUS
U.S. CITIES

City	State	Population	Facebook Followers	% of Population
Columbus	OH	905,748	24,000	2.65
Charlotte	NC	874,579	48,000	5.49
Greensboro	NC	299,035	30,000	10.03
Winston-Salem	NC	251,350	57,000	22.68
Port St. Lucie	FL	231,790	39,000	16.83
Birmingham	AL	200,733	43,000	21.42
Montgomery	AL	200,603	37,000	18.44
Akron	OH	190,469	36,000	18.90
Aurora	IL	177,856	75,000	42.17
Allentown	PA	125,845	33,000	26.22
Rock Hill	SC	75,048	20,000	26.65
Lancaster	PA	59,321	16,000	26.97
Charleston	WV	48,864	28,000	57.30
Charlottesville	VA	46,553	11,000	23.63
Huntington	WV	45,746	30,000	65.58
York	PA	44,800	11,000	24.55
Easton	PA	27,189	14,000	51.49
Union City	GA	26,409	3,500	13.25
Trotwood	OH	24,431	3,800	15.55
Athens	OH	20,820	6,800	32.66
Lenoir	NC	17,910	12,000	67.00
N. Myrtle Beach	SC	16,819	41,000	243.77
Live Oak	TX	15,781	14,000	88.71
Dumfries	VA	5,679	324	5.71
Total		3,933,378	633,424	16.10

In my research, Facebook is the most dominant social media platform being used by cities and towns. While large cities like Columbus, Ohio, and Charlotte, North Carolina have less than 6% of the population following their Facebook pages, the next 21 cities with populations between 299,035 and 15,781 have over 40% of their population as Facebook followers. While some of the followers may be former residents or family members who have a connection with the city, it is still a remarkable number of people to consume content and engage in government.

Strengths of Facebook for Local Government:

1. Wide Reach and Audience Engagement:
   - Facebook has a large user base, allowing local governments to reach a broad audience, including different age groups and demographics.
   - The platform's algorithm helps content reach people who are more likely to engage with it.

2. Cost-Effective Communication:
   - Facebook is a cost-effective way to communicate with residents compared to traditional media like print or TV.
   - Paid advertising options are relatively inexpensive and can be highly targeted.

3. Real-Time Interaction:
   - Local governments can provide real-time updates and announcements, essential for emergency communications and time-sensitive information.
   - The platform allows for immediate feedback and interaction with residents through comments, messages, and reactions.

4. Transparency and Accountability:
   - Regular updates and live streaming of meetings and events promote transparency.
   - Residents can see what the local government is doing and hold officials accountable.

5. Community Building:
   - Facebook groups and pages can foster a sense of community by allowing residents to share their concerns, ideas, and feedback.
   - It facilitates community engagement and participation in local government initiatives.

6. Event Promotion:
   - Local governments can promote events, public meetings, and community activities, increasing attendance and participation.
   - Facebook events can be shared easily, expanding their reach.

7. Data and Analytics:

- Facebook provides detailed analytics on post-performance, audience demographics, and engagement metrics.

- This data can be used to tailor communication strategies and improve outreach efforts.

Weaknesses of Facebook for Local Government:

1. Misinformation and Rumors:

- The platform can be a breeding ground for misinformation and rumors, which can spread quickly and be challenging to control.

- False information can damage public trust and create unnecessary panic or confusion.

2. Negative Feedback and Trolls:

- Public pages can attract negative comments, criticism, and trolling, which can be difficult to manage and moderate.

- Handling negative feedback requires a well-thought-out strategy to avoid escalating conflicts.

3. Algorithm Changes:

- Facebook frequently updates its algorithms, which can impact the visibility and reach of local government posts.

- This unpredictability makes it challenging to maintain consistent engagement.

4. Privacy Concerns:

- Facebook has faced numerous privacy scandals, which can affect public trust in the platform.

- Residents may be hesitant to engage with local government on a platform with perceived privacy risks.

5. Resource Intensive:

- Managing a Facebook presence effectively requires time, effort, and resources, including staff dedicated to social media management.

- Smaller local governments may struggle to allocate sufficient resources for this purpose.

6. Digital Divide:

- Not all residents use Facebook or have access to the internet, creating a digital divide.
- Relying solely on Facebook for communication can exclude certain segments of the population.

7. Content Overload:

- With the vast amount of content on Facebook, local government posts may get lost in users' feeds.
- Capturing and maintaining residents' attention requires compelling and frequent content updates.

8. Platform Dependence:

- Over-reliance on Facebook can be risky if the platform experiences technical issues or changes its policies.
- Diversifying communication channels is essential to ensure reliable outreach.

Local governments can leverage Facebook's strengths while addressing its weaknesses to optimize their social media strategy and effectively engage with their communities.

X (TWITTER) USE IN VARIOUS U.S. CITIES

City	State	Population	X (Twitter) Followers	% of Population
Columbus	OH	905,748	28,200	3.11
Charlotte	NC	874,579	211,500	24.18
Greensboro	NC	299,035	52,400	17.52
Winston-Salem	NC	251,350	24,300	9.67
Port St. Lucie	FL	231,790	9,011	3.89
Birmingham	AL	200,733	28,900	14.40
Montgomery	AL	200,603	10,700	5.33
Akron	OH	190,469	16,900	8.87
Aurora	IL	177,856	10,300	5.79
Allentown	PA	125,845	N/A	N/A
Rock Hill	SC	75,048	168	0.22
Lancaster	PA	59,321	6,736	11.36
Charleston	WV	48,864	6,677	13.66
Charlottesville	VA	46,553	18,300	39.31
Huntington	WV	45,746	1,619	3.54
York	PA	44,800	915	2.04
Easton	PA	27,189	1,901	6.99
Union City	GA	26,409	780	2.95
Trotwood	OH	24,431	0	0.00
Athens	OH	20,820	8,008	38.46
Lenoir	NC	17,910	750	4.19
N. Myrtle Beach	SC	16,819	21,100	125.45
Live Oak	TX	15,781	1,200	7.60
Dumfries	VA	5,679	1,469	25.87
Total		3,933,378	461,834	11.74

Twitter, now rebranded as X, is another significant player in social media use by cities and towns. While the difference in the percentage of the population from Facebook, is 16% Facebook to almost 12% for X, there is some volatility in the usage in cities. For instance, Charlottesville, Virginia has over 39% of the population as followers, while a similar-sized city, Huntington, West Virginia has less than 4% of the population as followers. Columbus, Ohio, and Charlotte, North Carolina have about 3% and 24% respectively while being nearly the same size population.

Strengths of X (Twitter) for Local Government:

1. Real-Time Communication:

   - X allows for instant updates and rapid dissemination of information, which is crucial during emergencies, weather alerts, or urgent announcements.

   - Local governments can provide live coverage of events and meetings, keeping residents informed in real-time.

2. Broad Reach and Engagement:

   - X has a large user base, including influential users and local media, which helps amplify messages.

   - The platform encourages interaction through replies, retweets, and likes, fostering engagement with the community.

3. Hashtags and Trends:

   - Hashtags make it easy to categorize content and join relevant conversations, increasing the visibility of posts.

   - Trending topics can help local governments tap into larger discussions and reach a broader audience.

4. Transparency and Accountability:

   - Regular updates and open communication on X can promote transparency in local government activities.

   - Residents can publicly ask questions and express concerns, holding officials accountable.

5. Direct Interaction:

   - X enables direct communication between local government officials and residents, facilitating a more personal connection.

   - Officials can respond to queries, clarify information, and engage in public dialogue.

6. Crisis Management:

   - The platform is effective for managing crises by providing timely updates and countering misinformation.

   - Quick responses can help mitigate the impact of crises and reassure the public.

7. Data and Analytics:

- X provides analytics on tweet performance, audience demographics, and engagement, helping local governments understand their reach and impact.

- Insights can inform future communication strategies and improve outreach efforts.

Weaknesses of X (Twitter) for Local Government:

1. Character Limit:

- The character limit (even after being increased) can restrict the amount of detailed information that can be shared in a single tweet.

- Complex issues may require multiple tweets or threads, which can be less effective for conveying information.

2. Misinformation and Fake News:

- The platform can facilitate the rapid spread of misinformation and rumors, which can be challenging to control.

- False information can damage public trust and create confusion or panic.

3. Negative Feedback and Trolls:

- X can attract negative comments, criticism, and trolling, which can be difficult to manage.

- Handling negative interactions requires careful moderation and a strategic approach.

4. Algorithm Changes:

- Frequent changes to the platform's algorithm can affect the visibility and reach of posts.

- This unpredictability makes it challenging to maintain consistent engagement.

5. Resource Intensive:

- Managing an effective presence on X requires time, effort, and dedicated resources, including staff for social media management.

- Smaller local governments may struggle to allocate sufficient resources for this purpose.

6. Digital Divide:

- Not all residents use X or have access to the internet, creating a digital divide.
- Relying heavily on X for communication can exclude certain segments of the population.

7. Ephemeral Nature:

- Tweets can quickly get buried in users' feeds due to the fast-paced nature of the platform.
- Important messages may be missed if not posted frequently or at optimal times.

8. Privacy and Security Concerns:

- X has faced issues with account security and privacy breaches, which can affect public trust.
- Residents may be hesitant to engage with local government on a platform with perceived security risks.

9. Short Attention Span:

- Users on X tend to have short attention spans due to the rapid flow of information.
- Capturing and maintaining residents' attention requires concise, compelling content.

10. Platform Dependence:

- Over-reliance on X can be risky if the platform experiences technical issues or changes its policies.
- Diversifying communication channels is essential to ensure reliable outreach.

Local governments can maximize the strengths of X by using it for real-time updates, engagement, and crisis management while addressing its weaknesses through strategic planning and diversification of communication channels.

YOUTUBE USE IN VARIOUS U.S. CITIES

City	State	Population	YouTube Subscribe	% of Population
Columbus	OH	905,748	7,310	0.81
Charlotte	NC	874,579	4,031	0.46
Greensboro	NC	299,035	4,990	1.67
Winston-Salem	NC	251,350	3,600	1.43
Port St. Lucie	FL	231,790	5,800	2.50
Birmingham	AL	200,733	2,170	1.08
Montgomery	AL	200,603	402	0.20
Akron	OH	190,469	1,400	0.74
Aurora	IL	177,856	1,470	0.83
Allentown	PA	125,845	73	0.06
Rock Hill	SC	75,048	1,100	1.47
Lancaster	PA	59,321	993	1.67
Charleston	WV	48,864	150	0.31
Charlottesville	VA	46,553	207	0.44
Huntington	WV	45,746	536	1.17
York	PA	44,800	N/A	N/A
Easton	PA	27,189	228	0.84
Union City	GA	26,409	177	0,67
Trotwood	OH	24,431	0	0.00
Athens	OH	20,820	579	2.78
Lenoir	NC	17,910	197	1.10
N. Myrtle Beach	SC	16,819	358	2.13
Live Oak	TX	15,781	15	0.10
Dumfries	VA	5,679	203	3.57
Total		**3,933,378**	**35,989**	**0.92**

YouTube is a powerful video-sharing platform that cities and towns can use to enhance communication and engagement with residents. Having a YouTube channel allows cities and towns to upload and share videos of public meetings, community events, and informational content, increasing accessibility and transparency.

The YouTube platform reaches a broad audience, providing visual and engaging content that can inform and educate residents. The live-streaming function allows local governments to interact directly with the community, building trust and promoting a sense of connection.

Strengths of YouTube for Local Government:

1. Wide Reach and Diverse Audience:

- YouTube has a vast user base, including diverse demographics, allowing local governments to reach a broad audience.
- The platform is accessible worldwide, which can help local governments connect with a larger community, including expatriates and global stakeholders.

2. Engaging Visual Content:

- Video content is highly engaging and can effectively convey complex information through visual and audio elements.
- Local governments can use videos to explain policies, showcase community events, and provide visual tours of projects and initiatives.

3. Educational and Informative:

- YouTube is an excellent platform for educational content, such as how-to guides, public service announcements, and informational videos.
- It allows local governments to educate residents on various topics, from recycling procedures to emergency preparedness.

4. Transparency and Accountability:

- Broadcasting council meetings, public hearings, and other official events live or as recorded videos, promote transparency.
- Residents can access these videos at their convenience, enhancing accountability and trust in local government operations.

5. Searchable and Archivable Content:

- YouTube videos are easily searchable and can be archived for future reference.
- This feature allows residents to access past content, providing a comprehensive library of information and updates.

6. Community Building:

- YouTube enables comments and discussions, fostering community engagement and dialogue.

- Local governments can interact with residents, address concerns, and gather feedback directly on the platform.

7. Monetization Opportunities:

- While not a primary goal, local governments can monetize their YouTube channels through ads, potentially generating revenue to support public projects.

- Partnerships and sponsorships can also provide additional funding opportunities.

8. Detailed Analytics:

- YouTube offers robust analytics, providing insights into viewer demographics, engagement metrics, and video performance.

- This data can help local governments refine their content strategy and improve outreach efforts.

Weaknesses of YouTube for Local Government:

1. Resource Intensive:

- Producing high-quality video content requires time, effort, and resources, including equipment and skilled personnel.

- Smaller local governments may struggle to allocate sufficient resources for video production and management.

2. Content Creation Challenges:

- Creating engaging and professional videos can be challenging, especially for complex or dry topics.

- Consistent content production requires creativity and ongoing effort to maintain viewer interest.

3. Misinformation and Negative Feedback:

- YouTube comments can include misinformation, negative feedback, and trolling, which can be challenging to manage.

- Moderating comments and addressing false information requires additional resources and strategic planning.

4. Algorithm Changes:

- YouTube frequently updates its algorithms, affecting video visibility and discoverability.

- Local governments may need to adapt their content strategy to stay visible and maintain engagement.

5. Digital Divide:

- Not all residents have access to high-speed internet or are comfortable using digital platforms, creating a digital divide.
- Relying heavily on YouTube for communication can exclude certain segments of the population.

6. Privacy Concerns:

- Privacy issues and data security concerns may deter some residents from engaging with content on YouTube.
- Local governments must ensure they handle user data responsibly and comply with privacy regulations.

7. Monetization and Ad Placement Issues:

- While monetization can be a benefit, ad placement may not always align with the values or messaging of local government content.
- Inappropriate or disruptive ads can detract from the viewing experience.

8. Content Saturation:

- With the vast amount of content on YouTube, local government videos may struggle to stand out and attract viewers.
- Effective promotion and optimization strategies are essential to reach the target audience.

9. Platform Dependence:

- Over-reliance on YouTube can be risky if the platform experiences technical issues or changes its policies.
- Diversifying communication channels ensures more reliable and widespread outreach.

Local governments can leverage YouTube's strengths by producing engaging, informative content and using the platform for transparency and community building while addressing its weaknesses through careful planning and resource allocation.

INSTAGRAM USE IN VARIOUS U.S. CITIES

City	State	Population	Instagram Followers	% of Population
Columbus	OH	905,748	10,800	1.19
Charlotte	NC	874,579	33,000	3.77
Greensboro	NC	299,035	8,280	2.77
Winston-Salem	NC	251,350	20,200	8.04
Port St. Lucie	FL	231,790	11,800	5.09
Birmingham	AL	200,733	19,900	9.91
Montgomery	AL	200,603	7,268	3.62
Akron	OH	190,469	2,708	1.42
Aurora	IL	177,856	15,200	8.55
Allentown	PA	125,845	3,366	2.67
Rock Hill	SC	75,048	8	0.01
Lancaster	PA	59,321	10,000	16.86
Charleston	WV	48,864	N/A	N/A
Charlottesville	VA	46,553	1,052	2.26
Huntington	WV	45,746	9,353	20.45
York	PA	44,800	3,199	7.14
Easton	PA	27,189	9,026	33.20
Union City	GA	26,409	2,087	7.90
Trotwood	OH	24,431	0	0.00
Athens	OH	20,820	2,648	12.72
Lenoir	NC	17,910	2,173	12.13
N. Myrtle Beach	SC	16,819	1,240	7.37
Live Oak	TX	15,781	805	5.10
Dumfries	VA	5,679	938	16.52
Total		3,933,378	175,057	4.45

Instagram is a social media platform that local governments can use to connect with their citizens through visually engaging content. While not as popular as Facebook and X in the local government sector, over 4% of the population of our sample group of U.S. cities are using the medium. It is very easy to use pictures and videos that you might also be using on other platforms. which allows you to repost content efficiently.

Strengths of Instagram for Local Government:

1. Visual Engagement:

- Instagram is highly visual, making it ideal for sharing photos and videos that can capture community events, infrastructure projects, and cityscapes.
- Visual content tends to engage audiences more effectively than text, making it easier to communicate important messages.

2. Stories and Reels:

- Instagram Stories and Reels offer dynamic ways to share real-time updates, behind-the-scenes looks, and short, engaging videos.
- These features can increase engagement and keep the community informed about daily activities and updates.

3. Audience Reach and Demographics:

- Instagram has a large and diverse user base, particularly among younger demographics, which can help local governments reach a broader audience.
- The platform's popularity among younger users makes it a valuable tool for engaging with this demographic.

4. Community Building and Engagement:

- Instagram encourages interaction through likes, comments, and direct messages, fostering community engagement.
- Local governments can use the platform to engage with residents, gather feedback, and promote community participation.

5. Hashtags and Discoverability:

- Hashtags can increase the discoverability of posts, allowing local governments to join larger conversations and reach a wider audience.
- Geotagging posts can also enhance discoverability among local users.

6. Transparency and Trust:

- Regular updates and visual transparency about government activities, projects, and events can build trust and accountability.

- Sharing positive stories and community successes can enhance the local government's reputation.

7. Analytics and Insights:

- Instagram provides analytics on post-performance, audience demographics, and engagement metrics.
- This data can inform content strategy and help local governments understand their audience better.

8. Cross-Platform Integration:

- Instagram integrates well with other social media platforms like Facebook, allowing for seamless cross-posting and content sharing.
- This integration can streamline social media management and amplify reach.

Weaknesses of Instagram for Local Government:

1. Resource Intensive:

- Creating high-quality visual content requires time, effort, and resources, including photography and videography skills.
- Regularly updating the platform and maintaining engagement can be challenging for smaller local governments with limited resources.

2. Algorithm Changes:

- Instagram's algorithm frequently changes, affecting the visibility and reach of posts.
- Local governments must stay updated with these changes and adapt their strategies accordingly.

3. Misinformation and Negative Feedback:

- The platform can be used to spread misinformation, and managing negative comments or misinformation can be resource intensive.
- Addressing false information and maintaining a positive image requires active moderation.

4. Short Attention Span:

- Users often scroll quickly through their feeds, so capturing and maintaining attention can be challenging.
- Content needs to be visually appealing and concise to be effective.

5. Privacy Concerns:

- Instagram has faced issues with privacy and data security, which can affect public trust.
- Residents may be hesitant to engage with local government on a platform with perceived privacy risks.

6. Digital Divide:

- Not all residents use Instagram or have access to the internet, creating a digital divide.
- Relying heavily on Instagram for communication can exclude certain segments of the population.

7. Limited Text and Link Options:

- Instagram's emphasis on visual content and limited text space can make it challenging to convey detailed information.
- Links are not clickable in regular posts, which can limit direct traffic to websites or detailed resources.

8. Ephemeral Content:

- Instagram Stories disappear after 24 hours, which can limit the longevity of important messages.
- Important updates may need to be reposted or highlighted to ensure they are seen by a broader audience.

9. Platform Dependence:

- Over-reliance on Instagram can be risky if the platform experiences technical issues or changes its policies.
- Diversifying communication channels ensures more reliable and widespread outreach.

Local governments can leverage Instagram's strengths by focusing on visually engaging content and community interaction while addressing its weaknesses through careful planning, resource allocation, and diversification of communication channels.

LINKEDIN USE IN VARIOUS
U.S. CITIES

City	State	Population	LinkedIn Followers	% of Population
Columbus	OH	905,748	350	0.04
Charlotte	NC	874,579	38,209	4.37
Greensboro	NC	299,035	9,000	3.01
Winston-Salem	NC	251,350	6,000	2.39
Port St. Lucie	FL	231,790	3,238	1.40
Birmingham	AL	200,733	N/A	N/A
Montgomery	AL	200,603	2,517	1.25
Akron	OH	190,469	1,096	0.58
Aurora	IL	177,856	4,021	2.26
Allentown	PA	125,845	3,000	2.38
Rock Hill	SC	75,048	2,598	3.46
Lancaster	PA	59,321	3,771	6.36
Charleston	WV	48,864	521	1.07
Charlottesville	VA	46,553	5,000	10.74
Huntington	WV	45,746	16	0.03
York	PA	44,800	393	0.88
Easton	PA	27,189	314	1.15
Union City	GA	26,409	1,013	3.84
Trotwood	OH	24,431	34	0.14
Athens	OH	20,820	0	0.00
Lenoir	NC	17,910	414	2.31
N. Myrtle Beach	SC	16,819	1,343	7.99
Live Oak	TX	15,781	450	2.85
Dumfries	VA	5,679	166	2.92
Total		3,933,378	83,464	2.12

While LinkedIn has the lowest percentage of the population of our sample set of cities at just above 2%, I believe it is going to play a role in the social media needs for cities and towns. Once viewed solely as a business networking platform, LinkedIn is becoming popular with city managers, senior staff and some elected officials. The platform's professional environment fosters a space for thoughtful dialogue and collaboration. It is a valuable economic development platform since it links the business community with the local government leaders and administrators. Few prospective businesses will be engaging with Facebook or X to see local economic development news. In my consulting work with cities, I find LinkedIn becoming a go to resource for city administrators and leaders.

Strengths of LinkedIn for Local Government:

1. Professional Audience:

- LinkedIn is widely used by professionals, making it an ideal platform for local governments to connect with business leaders, potential investors, and other government officials.

- The platform is well-suited for sharing detailed policy updates, economic development initiatives, and professional achievements.

2. Networking Opportunities:

- LinkedIn facilitates networking with other government agencies, non-profits, and private sector partners.

- Local governments can use the platform to build partnerships, collaborate on projects, and share best practices.

3. Talent Acquisition and Recruitment:

- LinkedIn is an effective tool for recruiting skilled professionals for government positions.

- Job postings and recruitment campaigns can reach a wide audience of qualified candidates.

4. Thought Leadership and Credibility:

- Local government officials can publish articles, share insights, and engage in discussions to establish themselves as thought leaders.

- Sharing expertise and success stories enhances the credibility and reputation of the local government.

5. Event Promotion and Participation:

- LinkedIn is useful for promoting conferences, workshops, and public meetings.

- The platform's event feature can help increase attendance and engagement from professionals and stakeholders.

6. B2B Communication:

- LinkedIn is effective for business-to-business (B2B) communication, making it a valuable platform for economic development and attracting businesses to the area.

- Sharing information about local business incentives, infrastructure projects, and economic trends can attract potential investors.

7. Analytics and Insights:

- LinkedIn provides detailed analytics on post-performance, audience demographics, and engagement metrics.
- This data can help local governments refine their communication strategies and improve outreach efforts.

8. Educational Content and Resources:

- The platform is suitable for sharing educational content, such as research reports, whitepapers, and policy documents.
- Local governments can use LinkedIn to educate the public and other stakeholders on important issues.

Weaknesses of LinkedIn for Local Government:

1. Narrow Audience:

- LinkedIn primarily attracts professionals and businesses, which may not represent the broader community.
- The platform may not be effective for reaching all segments of the population, especially those not in the professional or business sectors.

2. Less Engagement for General Public Issues:

- LinkedIn users are generally more interested in professional and industry-specific content, which may limit engagement on general public issues.
- Posts about community events, public services, or general announcements may receive less attention compared to professional content.

3. Content Creation Challenges:

- Creating content that resonates with LinkedIn's professional audience requires a different approach compared to other social media platforms.

- Local governments need to focus on professional, industry-relevant content, which may require additional resources and expertise.

4. Resource Intensive:

- Managing an effective presence on LinkedIn requires time, effort, and resources, including content creation and regular engagement.
- Smaller local governments may struggle to allocate sufficient resources for LinkedIn management.

5. Algorithm Changes:

- LinkedIn's algorithm changes can affect the visibility and reach of posts.
- Local governments need to stay updated with these changes and adapt their strategies accordingly.

6. Privacy Concerns:

- While LinkedIn is generally perceived as more professional, privacy concerns still exist, which can affect public trust.
- Local governments must ensure they handle user data responsibly and comply with privacy regulations.

7. Limited Multimedia Options:

- LinkedIn offers fewer multimedia options compared to platforms like Instagram or YouTube, which can limit creative content strategies.
- The platform's emphasis on text-based content may not suit all types of communication needs.

8. Platform Dependence:

- Over-reliance on LinkedIn can be risky if the platform experiences technical issues or changes its policies.
- Diversifying communication channels ensures more reliable and widespread outreach.

Local government administrators and elected officials can leverage LinkedIn's strengths by focusing on professional networking, recruitment, and thought leadership while addressing its weaknesses through strategic planning and resource allocation.

TIKTOK & SNAPCHAT IN VARIOUS U.S. CITIES

TikTok and Snapchat, while not traditional platforms for local government communication, offer unique opportunities to engage younger residents and share community events and updates creatively and informally. By creating short, engaging videos on TikTok, cities and towns can highlight local events, promote public safety tips, and showcase community initiatives in a format that resonates with a younger audience. Similarly, Snapchat's Stories and filters can be used to provide real-time updates, share behind-the-scenes glimpses of municipal operations, and encourage resident participation in local events. These platforms allow local governments to reach a broader audience, fostering a sense of community and civic engagement among younger residents who might not engage with more traditional forms of communication.

Future of Social Media in Local Government

The future of social media in local government will continue to evolve. New platforms will emerge, and cities need to stay current with these trends. Local governments need to dedicate time and resources to manage their social media presence effectively. This ongoing adaptation will ensure that cities can leverage social media to improve communication and engagement with their residents. The advancements in technology move at such a rapid pace, it is hard to envision what is coming, but entrepreneurial cities will be at the forefront of innovation and technology in social media.

Conclusion

Navigating the ever-evolving landscape of social media is both a challenge and an opportunity for local governments. As we've explored, each platform offers unique benefits, and while Facebook and YouTube emerge as the top contenders for cities and towns, it's crucial to remain adaptable. With nearly 60% of the global social media user base on Facebook, this platform provides an unparalleled reach for community engagement, live-streaming events, and disseminating vital information. YouTube, with its vast audience and ability to host long-form, searchable content, also presents significant opportunities for cities to enhance transparency and accessibility.

Yet, beyond these platforms, the future is less clear. TikTok's rising popularity must be weighed against its political uncertainties, and while X (formerly Twitter) has a strong presence in the U.S., its rebranding has caused confusion that may hinder its effectiveness for city communications. Meanwhile, LinkedIn, though currently underutilized by local governments, shows promise as a platform for engaging with city managers, staff, and elected officials.

The key takeaway for city leaders is that social media is no longer optional—it's a critical tool for modern governance. However, it's not just about being on every platform but strategically choosing the ones that align with your city's communication goals. Regularly evaluating your social media presence and staying informed about emerging trends will ensure your city remains connected and relevant.

Importantly, cities must remember that the true power of social media lies in its ability to humanize government, build trust, and fostering a sense of community. By sharing stories, celebrating local achievements, and engaging directly with residents, cities can create a more inclusive and vibrant community.

As we look to the future, it's also essential to acknowledge the digital divide. Ensuring that your social media strategy is inclusive and accessible to all residents will help bridge this gap and ensure that no one is left behind.

In conclusion, now is the time to act. Assess your current social media efforts, set clear and achievable goals, and take the necessary steps to engage your community in meaningful ways. The rewards of a well-executed social media strategy will be felt across your city for years to come, preparing you for the changes that are sure to come in this digital age

CONCLUSION: LEADING WITH ENTREPRENEURIAL VISION INTO THE FUTURE

> *"The end of a melody is not its goal: but nonetheless, had the melody not reached its end it would not have reached its goal either. A parable."*
> **-Friedrich Nietzsche**

Where do we go from here? Readers of a book like this often wonder why it was written and what solutions the author offers for the challenges discussed. After a lifetime working in the local government sector, I believe there is a significant transition coming throughout the globe based on factors that are already at play. Whether we call it the Entrepreneurial City, the Transitioning City, or the Reinvented City, we cannot continue to apply the same logic to a rapidly changing environment.

Take for example the story of UBER. Most companies wanting to compete against the taxi industry would commission lengthy studies and navigate complex regulatory processes. In contrast, Uber bypassed the traditional approach by launching their "ridesharing" service directly in a few cities, before taxi companies or regulators could react.

Some cities resisted, but Uber moved on, allowing public demand to push for wider acceptance. This disruptive strategy allowed them to transform the industry while the traditional taxi services, slow to adapt, were left behind. They couldn't adjust because their monopolistic service was caught flat-footed.

If cities and towns continue to support outmoded ways of looking at and solving problems, entrepreneurs will find a way to establish a foothold, and cities and towns will be on the sidelines, with their infrastructure and citizens being used as the laboratory for innovative services. Why not be in the leadership role and have some say in your future?

Cities face complex challenges that require innovative thinking to solve. Here are five major challenges and how an entrepreneurial approach can address them.

CLIMATE CHANGE AND ENVIRONMENTAL SUSTAINABILITY:

Regardless of political affiliation or philosophy, climate change is a pressing issue that impacts cities and towns in profound ways. The effects are widespread and costly, influencing many aspects of urban life. Cities are spending billions each year to manage these challenges and mitigate their impacts. Future growth, development, and the overall quality of life in urban areas are at stake, and it's important to recognize that there are no quick fixes to these complex problems.

Current Challenges:

✦ **Record Heatwaves and Severe Droughts**: Increasing temperatures and prolonged droughts strain water resources and infrastructure, affecting both daily life and long-term sustainability.

✦ **Devastating Wildfires**: Rising temperatures and changing weather patterns contribute to more frequent and severe wildfires, causing significant damage to property and ecosystems.

✦ **Coastal Erosion**: Rising sea levels and intensified storms lead to erosion and damage along coastlines, threatening coastal communities and infrastructure.

✦ **High Costs of Mitigation**: Managing and mitigating these environmental impacts require substantial financial investment, placing a strain on municipal budgets and resources.

Entrepreneurial Approach:

✦ **Innovation:** Entrepreneurs can drive the development of new green technologies and sustainable practices. For example, startups might create advanced energy-efficient systems, waste-to-energy technologies, or innovative solutions for carbon capture.

✦ **Resilience:** Entrepreneurs can foster adaptive strategies that incorporate sustainability into business models. Companies that focus on creating resilient infrastructure or offering eco-friendly products can lead by example.

✦ **Collaboration**: Partnerships between startups, local governments, and non-profits can accelerate the deployment of green solutions and raise awareness about sustainability.

Opportunity Application for Cities:

✦ **Solicit Clean Tech Startups**: Work to develop cooperative demonstration projects to prove feasibility. Manage risk and access federal, state, regional, and foundation support.

✦ **Transparent Operations Analysis**: Through an innovation office or designated staff person, invite analysis of operations through local colleges or universities and explore cutting-edge solutions. These exercises will expose some of the brightest minds in your area to embrace the challenge and come up with innovative solutions.

AFFORDABLE HOUSING AND INEQUALITY:

The affordable housing crisis is escalating and increasingly impacting communities across the country. This issue poses significant challenges and has far-reaching consequences for economic stability and social cohesion in cities and towns.

Current Challenges:

✦ **Lack of Affordable Workforce Housing**: The shortage of affordable housing jeopardizes the economic stability of many cities.

Essential public service workers, such as police officers and city employees, struggle to afford to live in the communities they serve, undermining the effectiveness and sustainability of these critical services.

✦ **Rising Homelessness and Economic Inequality**: The housing shortage contributes to increasing homelessness and exacerbates economic inequality. As more individuals are pushed to the margins, the social fabric of communities becomes more fragile, leading to greater instability and division.

✦ **Threat from Private Equity Firms**: The trend of private equity firms acquiring single-family homes threatens neighborhood stability. This practice drives up property prices, making homeownership increasingly unattainable for many, and potentially erodes the American Dream for a significant portion of the population.

Entrepreneurial Approach for Cities:

✦ Creative Solutions: Entrepreneurs can explore novel housing models such as modular homes, co-living spaces, or 3D-printed houses to address affordability.

✦ Community Involvement: Social enterprises can engage communities in housing development and management, ensuring that solutions are tailored to local needs.

✦ Scaling Innovations: Successful pilot projects can be scaled to address broader housing challenges, leveraging technology and innovative financing models.

Opportunity Application:

✦ **Affordable Housing Projects**: Create or support startups that develop cost-effective housing solutions or utilize innovative financing like community land trusts.

✦ **Impact Investment**: Invest in projects and companies focused on reducing inequality and providing affordable housing options.

TRANSPORTATION AND MOBILITY:

Transportation and mobility are increasingly critical issues for cities and towns as they navigate rapid urbanization and evolving technological landscapes. The rise of driverless cars, rideshare services, and pedestrian safety concerns are reshaping how we think about and manage transportation systems. At the same time, persistent problems like gridlock on highways contribute to significant economic and environmental costs.

Current Challenges:

+ **Traffic Congestion**: Gridlock on highways and major roads not only causes frustration for commuters but also results in billions of dollars in lost productivity each year. Congestion leads to increased pollution and higher accident rates, exacerbating the negative impacts on public health and urban environments.

+ **Increased Demand for Deliveries**: The surge in online shopping and delivery services has intensified pressure on already crowded streets and highways. Efficiently managing these delivery demands while avoiding further congestion is a significant challenge for modern cities.

+ **Pedestrian Safety**: Ensuring safe and accessible transportation options for pedestrians is crucial. Inadequate infrastructure can lead to increased accidents and injuries, affecting the overall quality of urban living.

Entrepreneurial Approach:

+ Disruptive Technology: Entrepreneurs can introduce disruptive innovations in transportation, such as autonomous vehicles, ride-sharing platforms, or electric bike rentals.

+ Smart Infrastructure: Develop and implement smart traffic management systems, and real-time transit information solutions to enhance urban mobility.

+ Scalable Solutions: Pilot new mobility solutions in smaller areas before scaling them city-wide, testing their efficacy and impact.

Opportunity Application for Cities:

+ Mobility-as-a-Service (MaaS): Develop platforms that integrate various transportation modes into a single service to improve accessibility and convenience.

✦ Urban Transport Startups: Invest in or start ventures focused on sustainable and efficient urban transport solutions.

PUBLIC HEALTH AND SAFETY:

Public health and safety are critical components of urban life that significantly impact both individual well-being and the broader functioning of cities and towns. The rising costs associated with healthcare, coupled with lost productivity due to health-related issues, represent a substantial burden on household budgets and municipal resources. Addressing these challenges requires innovative approaches to improve health outcomes and enhance overall safety within urban environments.

Current Challenges:

✦ **Healthcare Costs**: The cost of healthcare is a major expense for many households, affecting their financial stability and overall quality of life. High medical expenses can lead to financial strain and limit access to necessary services.

✦ **Productivity Losses**: Health issues contribute to absenteeism and decreased productivity among workers. When employees are unwell or dealing with health-related challenges, it impacts their performance and the delivery of essential public services.

✦ **Employee Morale**: An unhealthy workforce can lead to diminished employee morale and job satisfaction. High levels of stress and poor health can affect employee engagement and retention, further straining municipal resources.

✦ **Public Safety Concerns**: Ensuring public safety involves addressing a wide range of issues, from emergency response systems to crime prevention and disaster preparedness. Effective management of these areas is crucial for maintaining a secure and resilient urban environment.

Entrepreneurial Approach:

✦ Health Tech Innovation: Entrepreneurs can create new health technologies, such as telemedicine platforms, wearable health monitors, and smart emergency response systems.

✦ Preventive Measures: Develop startups that focus on health education, disease prevention, and wellness programs to address public health issues proactively.

✦ Data-Driven Solutions: Utilize big data and AI to predict and manage health crises, improving response times and resource allocation.

Opportunity Application for Cities:

✦ Health and Wellness Startups: Invest in or support companies that offer innovative health solutions or preventative care technologies.

✦ Smart Health Systems: Develop and deploy smart systems for tracking and responding to health and safety issues in urban environments.

TECHNOLOGICAL INTEGRATION AND CYBERSECURITY:

In an era of rapid technological advancement, cities face significant challenges in integrating new technologies while ensuring robust cybersecurity. As cities become increasingly reliant on digital systems, the risks associated with cyber threats and the necessity for seamless technological integration become more pronounced.

Current Challenges:

✦ **Cybersecurity Risks**: Cities are increasingly vulnerable to cyberattacks, including hacking and ransomware. These threats can disrupt critical services, compromise sensitive data, and erode public trust. The financial and reputational damage resulting from such attacks can be substantial, affecting everything from emergency response systems to municipal financial operations.

✦ **Cashless Transactions**: The growing trend towards cashless transactions means that more venues and businesses are accepting only credit and debit cards. Cities that fail to offer digital payment options for taxes, fees, and other services risk alienating a tech-savvy generation. This can lead to delays in payments, increased collection costs, and reduced citizen satisfaction.

Entrepreneurial Approach:

✦ **Cybersecurity Innovations**: Entrepreneurs can develop advanced cybersecurity solutions to protect urban infrastructure and data from cyber threats.

✦ **Tech Integration:** Create platforms that facilitate the integration of various technologies within the city, enhancing efficiency and connectivity.

✦ **Ethical Tech:** Promote startups that focus on ethical AI and technology use, ensuring that advancements do not exacerbate digital divides or privacy concerns.

Opportunity Application for Cities:

✦ **Cybersecurity Ventures**: Invest in or partner with companies that specialize in protecting urban digital infrastructure and sensitive data.

✦ **Smart City Platforms:** Develop and implement integrated technology solutions that improve urban services while ensuring robust cybersecurity measures.

I'm sure many are thinking, "This is an aggressive role for cities and towns. Where do we go for help?" I have listed several organizations in Appendix I that I believe can be of immense assistance. National organizations like the National League of Cities, the U.S. Conference of Mayors, and the International City/County Management Association are just a few of the many valuable resources available. On a state and regional level, the State Municipal Leagues and Regional Planning and Development Councils can act as the facilitators and repositories for what works and what doesn't work. The yearly conferences and regional meetings can put a spotlight on promising partnerships and innovative projects. Many of these organizations are striving to be more entrepreneurial in focus, but there is still much work to be done. As member-based organizations, get involved and advocate for information, speakers, panel discussions, and research on these important subjects.

These are exciting times, and the work you do on behalf of your city or town will echo for generations. Every person working in or representing cities and towns must rise to the challenge and approach each day with positivity and energy. Don't be afraid to reach out for help, and partner with those who bring innovation and new ideas. Technology has leveled the playing field, allowing even smaller cities and towns to lead in innovation and create places that draw residents and visitors with an exceptional quality of life.

I hope this book sparks inspiration and excitement. Like any book, this is just the beginning—the actions you take tomorrow and beyond will give it its true value. After a lifetime spent working on behalf of cities and towns, I deeply respect anyone who seeks to make things better. Be Amazing!

APPENDIX I

NATIONAL LEAGUE OF CITIES

660 N. Capitol St. NW
Washington, DC 20001
877-827-2385
Clarence Anthony, Executive Director
info@nlc.org

The National League of Cities (NLC) is an organization comprised of city, town, and village leaders who are focused on improving the quality of life for their current and future constituents.

With nearly 100 years of dedication to the strength and advancement of local governments, NLC has gained the trust and support of more than 2,700 cities across the nation. Our mission is to relentlessly advocate for, and protect the interests of, cities, towns, and villages by influencing federal policy, strengthening local leadership, and driving innovative solutions.

The National League of Cities is committed to a culture that values diversity and promotes inclusion and belonging. We are focused on building more inclusive communities and improving the quality of life for current and future residents. We strive to build teams at NLC that mirror our members and the communities they serve. Equity is a guiding principle for our work. In our mission to relentlessly advocate for and protect the interests of cities, towns, and villages, we believe in a fair distribution of resources based on a community's needs and a just approach that recognizes the impact of historic disinvestments in marginalized and underrepresented communities.

U.S. CONFERENCE OF MAYORS

1620 I St. NW
Washington, DC 20006
Tom Cochran, Executive Director
202-293-7330

The United States Conference of Mayors is the official non-partisan organization of cities with populations of 30,000 or more. There are over 1,400 such cities in the country today. Each city is represented in the Conference by its chief elected official, the mayor.

The Conference was born out of the Great Depression when in 1932, Detroit Mayor Frank Murphy invited the nation's mayors to his city to confront common problems caused by this dark time in our history.

Twenty-nine mayors gathered and, together, they called for Congress to provide relief, which Congress and the White House passed.

That first meeting galvanized the mayors to formalize their conference, and in February 1933 they did so in Washington. As it is today, the Conference continues to be the leading voice of cities in our nation's capital.

The Conference to this day is a nonpartisan forum where mayors engage directly with the President and Congress on the most pressing issues of the day.

INTERNATIONAL CITY/COUNTY MANAGEMENT ASSOCIATION

777 North Capitol St. NE
Suite 500
Washington, DC 20002-4201
Marc A. Ott, Executive Director
202-962-3680
800-745-8780

ICMA, from its inception in 1914, has continued to serve many essential functions for professional local government management executives. In so doing, the association has improved the quality of local government in which its members serve. As the number of local governments adopting the council-manager form of government has grown, so has the ICMA membership, in terms of both numbers and professional knowledge and skills.

In May 1991, members voted to change ICMA's constitutional name to the International City/County Management Association, reflecting the membership of the association and recognizing the evolution in county government professionalism.

Our 13,000+ members serve and improve lives for communities, from small towns with populations of a few hundred to metropolitan areas with populations of millions. ICMA offers professional development programs, research, publications, data and information, technical assistance, and training to create excellence in local governance and foster professional local government management worldwide.

NATIONAL ASSOCIATION OF REGIONAL COUNCILS

660 N. Capitol St. NW
Washington, DC 20001
Erich Zimmermann, Executive Director
erich@narc.org
202-986-1032

The National Association of Regional Councils (NARC) serves as the national voice for regions by advocating for regional cooperation as the most effective way to address a variety of community planning and development opportunities and issues.

NARC members include regional councils, councils of governments (COGs), regional planning and development agencies, Metropolitan Planning Organizations (MPOs), and other regional organizations. Members work collaboratively with their communities – large and small, urban and rural – to address their citizens' needs and promote a regional approach to planning for the future.

NARC member organizations are driven by a dedicated group of local elected officials and professionals who work with community leaders and citizens to build local capacity; develop and implement strategic investment plans; foster regional cooperation and economic competitiveness; forge public and private partnerships; and secure and manage funds from local, state, federal, and private sources.

INTERNATIONAL TOWN-GOWN ASSOCIATION

1250 Tiger Boulevard
Clemson, SC 29631
Beth Bagwell, MPA, Executive Director
beth@itga.org
864-616-2987

The International Town and Gown Association is a global non-profit association dedicated to college campus and community interests. To fulfill the mission and to meet the individual and community needs of its members, ITGA convenes gatherings, educates members and partner organizations, researches topics of concern, shares knowledge, and mobilizes members to ensure the success of communities around the world.

We are a diverse group of college and university administrators, city officials, and campus neighbors who collaborate on issues of campus and community. Economic development, sustainability, student housing, diversity and inclusion, quality of life, and civic engagement are among the many topics that ITGA addresses. Through its knowledgeable staff and professionally accomplished members, ITGA helps campuses and communities stay on top of current events.

ITGA helps its members in a number of ways, including an annual conference, networking and professional advice, regional gatherings, and collaborative problem-solving.

HAVEN Creative®

1300 South Blvd. Suite B42
Charlotte, NC 28203
Jeni Bukolt, Founder & CEO
hello@havencreativeagency.com
(704) 885-1857

On one of the most listened-to episodes of the Amazing Cities and Towns Podcasts, I interviewed a dynamic entrepreneur named Jeni Bukolt from HAVEN Creative®. Jeni understands the importance of branding and has done some amazing work with a growing number of entrepreneurial cities throughout the country.

HAVEN Creative® is a communications consultancy that helps organizations grow through brand positioning, PR, and advertising.

As a local government leader, you know you need to gain awareness for your organization, align your team, and prove the results, but how? Which marketing tactic makes sense for your brand, how do you tell a cohesive story, and where should you spend your time and effort?

If you are looking for a firm that understands local government and has a portfolio of successful rebranding projects with various-sized cities, look at their website at www.havencreativeagency.com. They have many free resources and examples of their work.

CiviSocial

www.civisocial.com
Sam Toles, Founder and CEO
sam@civisocial.com

CiviSocial offers a simple, 2-week engagement to identify the hidden future heroes on your team and give them the tools to successfully tell your story. Within your existing staff is the key to successful social communication, your own hidden storytellers. We identify them and give them the right tools and tactics.

Sam Toles, the founder of CiviSocial led the Vimeo platform, was Chief Content Officer for Bleacher Report, ran Digital for MGM Studios, and most recently worked with Mr. Beast's team. Noone has seen more success in social storytelling.

You have critical things to do and limited resources. That's why we created a program for small-to-midsized civic agencies, tailor made for their needs and budget realities. We are not a marketing agency, nor are we a creative agency. We aren't looking to sell you these services or do this work for you because it is both costly and unnecessary. You already have the resources on your own team to be successful.

STATE MUNICIPAL LEAGUES

State municipal leagues are excellent sources of information and provide educational sessions at their annual conferences and regional workshops.

Alaska Municipal League

(www.akml.org)
217 Second St Ste 200
Juneau, AK 99801
Nils Andreassen, Executive Director
(907) 586-1325
nils@akml.org

Alabama League of Municipalities

(www.almonline.org)
535 Adams Ave
Montgomery, AL 36104-4333
Greg Cochran, Executive Director
(334) 262-2566
gcochran@almonline.org

Arkansas Municipal League

(www.armunileague.org)
P. O. Box 38
North Little Rock, AR 72115Mark R. Hayes, Executive Director
(501) 374-3484 (218)
mhayes@arml.org

League of Arizona Cities and Towns

(www.azleague.org)
1820 West Washington Street
Phoenix, AZ 85007Thomas Belshe, Executive Director
(602) 258-5786
tbelshe@azleague.org

League of California Cities

(www.calcities.org)
1400 K St Fl 4Th
Sacramento, CA 95814-3916
Carolyn Coleman, CEO & Executive Director
(916) 658-8200
ccoleman@cacities.org

Colorado Municipal League

(www.cml.org)
1144 Sherman Street
Denver, CO 80203Kevin Bommer, Executive Director
(303) 831-6411
kbommer@cml.org

Connecticut Conference of Municipalities

(www.ccm-ct.org)
545 Long Wharf Drive
New Haven, CT 6511
Joe Allen DeLong, Executive Director
(203) 498-3000
jdelong@ccm-ct.org

Delaware League of Local Governments

PO Box 484
Dover, DE 19903-0484
Kevin Spence, Executive Director
(302) 678-0991
deleaguelocalgovs@gmail.com

Florida League of Cities Inc

(www.flcities.com)
301 S Bronough St Ste 300
Tallahassee, FL 32301-1722
Jeannie Garner, Executive Director/CEO
(850) 222-9684
jgarner@flcities.com

Georgia Municipal Association

(www.gacities.com)
201 Pryor St SW
Atlanta, GA 30303-3606
Larry Hanson, CEO & Executive Director
(404) 688-0472
lhanson@gacities.com

Iowa League of Cities

(www.iowaleague.org)
500 SW 7th Street, Suite 101
Des Moines, IA 50309
Alan Kemp, Executive Director
(515) 244-7282
alankemp@iowaleague.org

Association of Idaho Cities

(www.idahocities.org)
3100 S Vista Ave Ste 310
Boise, ID 83705-7346
Kelley Packer, Executive Director
(208) 344-8594
kpacker@idahocities.org

Illinois Municipal League

(www.iml.org)
500 East Capitol Avenue
Springfield, IL 62701Brad Cole, Executive Director
(217) 525-1220
bcole@iml.org

Accelerate Indiana Municipalities

(www.aimindiana.org)
125 W Market St St 100
Indianapolis, IN 46204Matthew Greller, Chief Executive Officer
(317) 237-6200
mgreller@aimindiana.org

League of Kansas Municipalities

(www.lkm.org)
300 S.W. 8th Avenue
Topeka, KS 66603-3951
Nathan Eberline, Executive Director
(785) 354-9565
neberline@lkm.org

Kentucky League of Cities

(www.klc.org)
100 E. Vine Street
Lexington, KY 40507J.D. Chaney, Executive Director/CEO
(859) 977-3700
jchaney@klc.org

Louisiana Municipal Association

(www.lma.org)
6767 Perkins Road
Baton Rouge, LA 70808
Richard Williams, Interim Executive Director
225-344-5001
rwilliams@lma.org

Massachusetts Municipal Association

(www.mma.org)
3 Center Plaza
Boston, MA 02108
Adam Chapdelaine, Executive Director & CEO
(617) 426-7272
achapdelaine@mma.org

Maryland Municipal League

(www.mdmunicipal.org)
1212 West Street
Annapolis, MD 21401Theresa Kuhns, CEO
(410) 295-9100
TheresaK@mdmunicipal.org

Maine Municipal Association
(www.memun.org)
60 Community Dr
Augusta, ME 04330-8008
Cathy Conlow, Executive Director
(207) 623-8428
cconlow@memun.org

Michigan Municipal League
(www.mml.org)
1675 Green Road
Ann Arbor, MI 48105-2530
Daniel P. Gilmartin, Executive Director & CEO
(734) 662-3246
dpg@mml.org

League of Minnesota Cities
(www.lmc.org)
145 University Avenue, West
St. Paul, MN 55103-2044
Luke Fischer, Executive Director
(651) 281-1200
lfischer@lmc.org

Missouri Municipal League
(www.mocities.com)
1727 Southridge Drive
Jefferson City, MO 65109
Richard Sheets, Executive Director
(573) 635-9134
rsheets@mocities.com

Mississippi Municipal League
(www.mmlonline.com)
600 E Amite St Ste 104
Jackson, MS 39201-2807
Shari T. Veazey, Executive Director
(601) 353-5854
shari1@mmlonline.com

Montana League of Cities and Towns

(www.mtleague.org)
208 North Montana – Suite 201
Helena, MT 59601-3837
Kelly Lynch, Executive Director
(406) 442-8768
kelly.lynch@mtleague.org

North Carolina League of Municipalities

(www.nclm.org)
434 Fayetteville Street
Raleigh, NC 27601
Rose Vaughn Williams, Executive Director
(919) 715-4000
rwilliams@nclm.org

North Dakota League of Cities

(www.ndlc.org)
410 East Front Avenue
Bismarck, ND 58504-5461
Matthew Gardner, Executive Director
(701) 223-3518
matt@ndlc.org

League of Nebraska Municipalities

(www.lonm.org)
1335 L St
Lincoln, NE 68508
L. Lynn Rex, Executive Director
(402) 476-2829
lynnr@lonm.org

New Hampshire Municipal Association

(www.nhmunicipal.org)
25 Triangle Park Drive
Concord, NH 3301
Margaret Byrnes, Executive Director
(603) 224-7447
mbyrnes@nhmunicipal.org

New Jersey State League of Municipalities

(www.njlm.org)
222 West State Street
Trenton, NJ 8608Mike Cerra, Executive Director
(609) 695-3481
mcerra@njlm.org

New Mexico Municipal League

(www.nmml.org)
PO Box 846
Santa Fe, NM 87504-0846
AJ Forte, Executive Director
(505) 982-5573
ajforte@nmml.org

Nevada League of Cities and Municipalities

(www.nvleague.org)
201 N. Carson Street, Suite 2
Carson City, NV 89701
Glenn Leavitt, Executive Director
(702)738-2128
gleavitt@nvleague.org

New York State Conference of Mayors and Municipal Officials

(www.nycom.org)
119 Washington Avenue
Albany, NY 12210
Barbara VanEpps, Executive Director
518-463-1185
barbara@nycom.org

Ohio Municipal League

(www.omlohio.org)
175 South Third Street, Suite 510
Columbus, OH 43215
Kent Scarrett, Executive Director
(614) 221-4349
kscarrett@omlohio.org

Oklahoma Municipal League Inc

(www.oml.org)
201 North East 23rd St
Oklahoma City, OK 73105
Mike Fina, Executive Director
(405) 528-7515
mfina@oml.org

League of Oregon Cities

(www.orcities.org)
1201 Court St NE 200
Salem, OR 97301
Patty Mulvihill, Executive Director
(503) 588-6550
pmulvihill@orcities.org

Pennsylvania Municipal League

(www.pml.org)
414 North Second Street
Harrisburg, PA 17101
John Brenner, Executive Director
(717) 236-9469
jbrenner@pml.org

Rhode Island League of Cities and Towns

(www.rileague.org)
1 State Street Ste 502
Providence, RI 2908
Ernie Almonte, Executive Director
(401) 272-3434
ealmonte@rileague.org

Municipal Association of South Carolina

(www.masc.sc)
1411 Gervais St
Columbia, SC 29201-3342
Todd Glover, Executive Director
(803) 799-9574
TGlover@masc.sc

South Dakota Municipal League

(www.sdmunicipalleague.org)
208 Island Drive
Fort Pierre, SD 57532
David Reiss, Executive Director
(605) 224-8654
David@sdmunicipalleague.org

Tennessee Municipal League

(www.tml1.org)
226 Anne Dallas Dudley Blvd., Suite 710
Nashville, TN 37219-1894
Anthony Haynes, Executive Director
(615) 255-6416
ahaynes@tml1.org

Texas Municipal League

(www.tml.org)
1821 Rutherford Lane Ste 400
Austin, TX 78754-5128
Bennett Sandlin, Executive Director
512-231-7400
exec@tml.org

Utah League of Cities and Towns

(www.ulct.org)
50 South 600 East, Suite 150
Salt Lake City, UT 84102
Cameron Diehl, Executive Director
(801) 328-1601
cdiehl@ulct.org

Virginia Municipal League

(www.vml.org)
13 E. Franklin Street
Richmond, VA 23219
Michelle Gowdy, Executive Director
(804) 649-8471
mgowdy@vml.org

Vermont League of Cities and Towns

(www.vlct.org)
89 Main Street, Suite 4
Montpelier, VT 05602-2948
Ted Brady, Executive Director
(802) 229-9111
tbrady@vlct.org

Association of Washington Cities

(www.wacities.org)
1076 Franklin St SE
Olympia, WA 98501
Deanna Dawson, CEO
(360) 753-4137
deannad@awcnet.org

League of Wisconsin Municipalities

(www.lwm-info.org)
316 W. Washington Ave., Suite 600
Madison, WI 53703
Jerry Deschane, Executive Director
(608) 267-2380
jdeschane@lwm-info.org

West Virginia Municipal League

(www.wvml.org)
2020 Kanawha Blvd. East
Charleston, WV 25311
Susan Economou, Executive Director
(304) 342-5564
seconomou@wvml.org

Wyoming Association of Municipalities

(www.wyomuni.org)
315 West 27th Street
Cheyenne, WY 82001
J. David Fraser, Executive Director
(307) 632-0398
dfraser@wyomuni.org

APPENDIX II

The Amazing Cities and Towns Podcast

Launched in April 2020, "The Amazing Cities and Towns Podcast" has quickly grown into a vital resource for local government leaders, amassing over 125 episodes to date. Producing this podcast has been a labor of love for the team at Amazing Cities, and I am profoundly grateful to the incredible guests who have generously shared their time and expertise. The podcast serves as a dynamic platform for the exchange of innovative ideas and practical solutions, featuring thought leaders from around the world who are shaping the future of our cities and towns.

We've had the privilege of hosting a wide range of experts, including New York Times bestselling author Sam Quinones, who provided a deep understanding of the opioid, methamphetamine, and fentanyl crises affecting communities nationwide and Jeni Bukolt of HAVEN Creative, who offered invaluable insights into city branding. These episodes, among many others, have contributed to building an impressive repository of knowledge, making the podcast a go-to resource for city officials, planners, and community leaders seeking guidance and inspiration.

When I first started the podcast, I didn't anticipate its longevity or the extent of its impact. However, that changed when a young council member approached me at a National League of Cities conference and shared how he regularly listens to the podcast and applies what he learns in his council work. This encounter illuminated the podcast's true value, extending far beyond the airing of each episode and providing ongoing support to those in local government.

The Amazing Cities and Towns Podcast continues to receive interview requests and inquiries from listeners and media outlets, reflecting its relevance and influence over the past four years. As it evolves, the podcast remains committed to delivering timely and actionable information that empowers city leaders to address the challenges and opportunities facing their communities.

Please review the following topics and visit the Amazing Cities website (www.amazingcities.org) and click on the Podcast tab to listen to any of the episodes.

Episode 1:

The Amazing Cities & Towns Podcast with Jim Hunt. Jim introduces the podcast and discusses his book, **The Amazing City-7 Steps to Creating an Amazing City**.

Episode 2:

The Amazing Cities & Towns Podcast with Jim Hunt talks about his book, The Amazing City-7 Steps to Creating an Amazing City. This episode talks about **ATTITUDE** and the importance of having a positive attitude in local government.

Episode 3:

The Amazing Cities & Towns Podcast with Jim Hunt talks about his book, The Amazing City-7 Steps to Creating an Amazing City. This episode talks about **MOTIVATION** and helping you find the keys to motivating yourself and others to create Amazing Cities.

Episode 4:

The Amazing Cities & Towns Podcast with Jim Hunt talks about his book, The Amazing City-7 Steps to Creating an Amazing City. This episode talks about **ATTENTION TO DETAIL** a critical component to building an Amazing city.

Episode 5:

The Amazing Cities & Towns Podcast with Jim Hunt talks about his book, The Amazing City-7 Steps to Creating an Amazing City. This episode talks about **ZING** and how to find or create the singular that can help define your Amazing City.

Episode 6:

The Amazing Cities & Towns Podcast with Jim Hunt talks about his book, The Amazing City-7 Steps to Creating an Amazing City. This episode talks about **INCLUSIVENESS** and the importance of building an inclusive environment.

Episode 7:

The Amazing Cities & Towns Podcast with Jim Hunt talks about his book, The Amazing City-7 Steps to Creating an Amazing City. This episode talks about **NEIGHBORHOOD EMPOWERMENT** and remembering that neighborhoods are the core for building an Amazing City.

Episode 8:

The Amazing Cities & Towns Podcast with Jim Hunt talks about his book, The Amazing City-7 Steps to Creating an Amazing City. This episode talks about **GREEN AWARENESS** and how to build sustainable cities and the benefits to the citizens and the community in general.

Episode 9:

The Amazing Cities & Towns Podcast with Jim Hunt. This episode talks about podcasting and the value of using a podcast to build a brand for your city or organization and the steps you need to use to start one.

Episode 10:

The Amazing Cities & Towns Podcast with Jim Hunt discusses the benefit of **State Municipal Leagues** using podcasts to highlight important issues facing cities and towns in their respective states.

Episode 11:

The Amazing Cities & Towns Podcast with Jim Hunt welcomes **Joe Moore, former Alderman in Chicago, Illinois** for 28 years. Joe and Jim talk about the Amazing city of Chicago and how local government works in one of the largest cities in the United States. A great discussion about the many neighborhoods that make up the city and how there is a small town feel in this huge city.

Episode 12:

Amazing Cities & Towns Podcast with Jim Hunt, Host welcomes **Lisa Powell-Graham, a motivational speaker and organizer or empowering retreats for women.** Lisa talks about her life's journey to empower women and how she got involved in national politics by working for the Hilary Clinton campaign for President and the lessons she learned. She is also a TEDx speaker and she explains how she prepared and delivered her TEDx talk.

Episode 13:

The Amazing Cities & Towns Podcast with Jim Hunt welcomes **Dave McCauley, the Mayor of Buckhannon, West Virginia** who talks about revitalization and community development. The City of Buckhannon has undergone a remarkable transformation and Mayor McCauley talks about how the community came together to enact change.

Episode 14:

The Amazing Cities & Towns Podcast with Jim Hunt welcomes **Mike Conduff, former ICMA Board Member and city manager in major Texas and Kansas cities**. Mike and Jim discuss improving City Council-City Manager Relations. Mike gives an insight into how city managers view their city councils and strategies to work together without conflict.

Episode 15:

The Amazing Cities & Towns Podcast with Jim Hunt welcomes **Kathie Novak, former Mayor of North Glenn, Colorado and Past President of the National League of Cities**. Kathie is a well-respected speaker and consultant who works with city councils throughout the United States to improve communication and effectiveness.

Episode 16:

The Amazing Cities & Towns Podcast with Jim Hunt recaps **the 2020 NLC Congressional Cities Conference** and coincidentally the beginning of the COVID-19 pandemic for cities and towns across the United States. This piece of history was the talk of the conference with masks and hand sanitizer at every session and event. A candid recollection at the beginning of a sea change in the lives of local government.

Episode 17:

The Amazing Cities & Towns Podcast with Jim Hunt welcomes Bobby Monroe and Kandi Marler of DataMax Corporation of Winston-Salem, North Carolina who discuss their company's efforts to provide revenue enhancement services for cities and towns. DataMax specializes in identifying and collecting business taxes on behalf of cities and towns.

Episode 18:

The Amazing Cities & Towns Podcast with Jim Hunt talks about the spreading pandemic and how cities and towns can prepare and manage this crisis.

Episode 19:

The Amazing Cities & Towns Podcast with Jim Hunt **Wayne Worth, community volunteer in West Virginia**. Wayne talks with Jim about his life of volunteering and what drives him to spend thousands of hours each year working on behalf of individuals and communities. The information shared can be a blueprint for cities and towns to identify those community "sparkplugs" who can facilitate change.

Episode 20:

The Amazing Cities & Towns Podcast with Jim Hunt welcomes **Vince Williams, Mayor of Union City, Georgia** to discuss how he has created a shared vision to build his city into a film and television location that is gaining national attention. Vince talks about how to recover from a major business loss in your community and to build relationships to get through tough times.

Episode 21:

The Amazing Cities & Towns Podcast with Jim Hunt welcomes **Melodee Colbert-Kean, a council member from Joplin, Missouri** about recovering from a major disaster. Joplin was hit with a devasting tornado, with a major loss of life and property. A compelling episode about what to do when the unforeseeable hits.

Episode 22:

The Amazing Cities & Towns Podcast with Jim Hunt with show producer, Matt James about the value of podcasting during a crisis, such as the Covid pandemic. Jim relates his experiences and gives details of the operation of a podcast and why it might be an option for cities and towns, as a communication tool.

Episode 23:

The Amazing Cities & Towns Podcast with Jim Hunt with show producer Matt James discusses the technical side of how a podcast works and what goes into editing and producing a professional podcast. An interesting look behind the curtain!

Episode 24:

The Amazing Cities & Towns Podcast with Jim Hunt **with Matt James, producer for the Amazing Cities and Towns Podcast** discusses social media and podcasting as tools for local governments. Matt gives insightful advice about cities and towns using social media and why a podcast might be a good complement to the toxic nature of social media.

Episode 25:

The Amazing Cities & Towns Podcast with Jim Hunt concludes his 4-part series on Podcasting for Local Governments and reviews the strengths and weaknesses for cities and towns starting a podcast. Needed advice from an expert in the field.

Episode 26:

The Amazing Cities & Towns Podcast with Jim Hunt interviews **Darryl Moss, the first African American Mayor of Creedmoor, North Carolina** about the differences between being mayor in one of the largest cities in the United States to the trials and tribulations of leading a small town in rural North Carolina. A great discussion to bring into perspective how small towns matter.

Episode 27:

The Amazing Cities & Towns Podcast with Jim Hunt recaps the first twenty-six episodes and talks about lessons learned and some of the mistakes and mishaps that occur when producing a podcast.

Episode 28:

The Amazing Cities & Towns Podcast with Jim Hunt. This episode is titled "City Finances in a Crisis" and features **Cathy Spain, a partner with Bearing Advisors and the former Chief Lobbyist for the Government Finance Officers Association (GFOA).** Cathy discusses the important factors to consider when a crisis arises. An important episode!

Episode 29:

The Amazing Cities & Towns Podcast with Jim Hunt. This episode features **Mike Conduff, a former Board Member of the ICMA and veteran city manager** who discusses how to recover as a city when crisis strikes. Mike is an acclaimed author of over 20 books on local governance.

Episode 30:

The Amazing Cities & Towns Podcast with Jim Hunt and **Joe Buscaino, Los Angeles City Councilman and the 2020 President of the National League of Cities** discusses how his city is coping during the COVID-19 crisis. Joe also talks about how the National League of Cities is serving as a resource and clearing house for cities and towns during the current challenges of a worldwide pandemic.

Episode 30:

The Amazing Cities & Towns Podcast with Jim Hunt. This episode, "How to Build a Brand for Your City or Town" talks with **Jeni Bukolt, Founder of Haven Creative, a branding and communication firm in Charlotte, North Carolina**. Jeni talks about the process for a city or town to follow when developing the brand for your community.

Episode 31:

The Amazing Cities & Towns Podcast with Jim Hunt discusses the effective use of social media by local governments. The guest is **Agnes Queen, a County Commissioner in Lewis County, West Virginia**. The episode is titled, "Local Government's Social Media Queen" and is a fun conversation with someone who understands social media and is using it effectively. A must listen!

Episode 32:

The Amazing Cities & Towns Podcast with Jim Hunt talks about "The Black-Tie Experience for Cities and Towns" with **Bob Pacanovsky, a customer experience expert.** Bob works with businesses and wants to talk about how cities and towns can benefit from secrets that businesses use to enhance their brand and profits. A remarkable insight from someone who knows how to give that Black Tie service.

Episode 33:

The Amazing Cities & Towns Podcast with Jim Hunt talks with the **Editor of the West Virginia Living Magazine, Nikki Bowman Mills** about how she leverages her media platforms to champion communities, build better businesses, and inspire action. She discusses a campaign called, "Turn This Town Around" and other ideas to build Amazing communities.

Episode 34:

The Amazing Cities & Towns Podcast with Jim Hunt welcomes **Anjelica Scott, a young African American community activist** in Clarksburg, West Virginia. Jim discusses with Anjelica a controversial issue of a Civil War statue in front of her county's courthouse. An interesting discussion from the perspective of the younger generation.

Episode 35:

The Amazing Cities & Towns Podcast with Jim Hunt welcomes **Ms. Ashley Shiwarski, Director of Business Development for HomeServe USA, and Eric Maxon of Stewards Plumbing in Albuquerque, New Mexico** for a discussion on how the remote work environment brought about because of COVID-19 has put a strain on residential plumbing. Helpful information for homeowners not wanting to experience a sewer line failure which can cost thousands of dollars.

Episode 36:

The Amazing Cities & Towns Podcast with Jim Hunt welcomes **Ryan Tolley, Director of the Robinson Grand Performing Arts Center in Clarksburg, West Virginia** to discuss how the pandemic has impacted his organization. Ryan talks about the innovative ways that he is working to provide quality events and maintain community interest in a time when group concerts and events are discouraged or prohibited.

Episode 37:

The Amazing Cities & Towns Podcast with Jim Hunt features **Stephen Morris, a local government consultant from Montgomery, Alabama.** Jim and Stephen have an interesting discussion about local government in Alabama and how his company has been able to help local counties and cities with revenue enhancement services that brings needed dollars back into communities from companies that have neglected to obtain business licenses and pay their business taxes.

Episode 38:

The Amazing Cities & Towns Podcast with Jim Hunt welcomes **Scott Hancock, Executive Director of the Maryland Municipal League** to the show to discuss how his league is operating during the pandemic and the overall value of state municipal leagues in general. Scott has served in multiple roles within local government, starting as a police officer in a small Maryland community and as a city manager which gives him a great perspective for his current role.

Episode 39:

The Amazing Cities & Towns Podcast with Jim Hunt discusses the Homelessness issue throughout the United States with **Jim Brooks, Senior Director of the National League of Cities.** Jim has had a variety of roles with NLC and is one of the most knowledgeable people on community development and housing in the country. Jim's insights are always spot on and he discusses the increasing homeless populations in American cities, as well as worldwide.

Episode 40:

The Amazing Cities & Towns Podcast with Jim Hunt and his guest on today's episode is **Mark McMillion, a leadership consultant** who graduated from the United States Military Academy at West Point and spent 22 years as an active-duty Army Officer. Mark has lived throughout the United States and the world during his Army career and shares some interesting lessons about community and leadership.

Episode 41:

The Amazing Cities & Towns Podcast with Jim Hunt welcomes **Phil Riley, Bearing Advisors Managing Director** to discuss the local government work of his company. Phil has a long relationship with the National League of Cities through his company's work with the National League of Cities Service Line Warranty Program. With a long career in the natural gas business, Phil on customer service and the importance of good Public/Private Partnerships.

Episode 42:

The Amazing Cities & Towns Podcast with Jim Hunt. Jim welcomes **Cindy Stewart, former Vice-President of the International Council of Shopping Centers for over 20 years and now a consultant to cities and towns on retail development.** Cindy talks about the process that companies go through to locate in a community. Her insights are valuable to any city or town interested in effective economic development.

Episode 43:

The Amazing Cities & Towns Podcast with Jim Hunt. **Cliff Johnson, Executive Director of the National League of Cities, Youth, Education, and Families Institute** discusses the 20th anniversary of the Institute and the incredible work that it has done. Cliff describes how the Youth, Education, and Families Institute has become a go-to place for Mayors, city managers, and others in Local Government.

Episode 44:

The Amazing Cities & Towns Podcast with Jim Hunt. **Ariel Guerrero joins Jim to discuss his work as a Racial Equity Practitioner and Consultant** working to dismantle structural and institutional racism through policy, procedure, and practice (3P's). Ariel had a successful career at the National League of Cities before entering his current position. A frank and in-depth discussion on a difficult subject.

Episode 45:

The Amazing Cities & Towns Podcast with Jim Hunt. The Amazing Cities Podcast producer Matt James takes the microphone and interviews host Jim Hunt about the first year of podcasting for Amazing Cities. They review notable guests and the overall impact of having a podcast dedicated to local government. Matt also goes into some of the interesting parts of producing a podcast.

Episode 46:

The Amazing Cities & Towns Podcast with Jim Hunt. Matt James, producer of the Amazing Cities Podcast returns to discuss the upcoming year with Jim Hunt, Host of the Amazing Cities Podcast. With a presidential inauguration and the recovery from the pandemic, 2021 looks to be an exciting year for local government. The Infrastructure Bill will put over a trillion dollars' worth of bridges, roads, and broadband into the country and local governments will be an integral part of the process.

Episode 47:

The Amazing Cities & Towns Podcast with Jim Hunt welcomes **Mark Houser, Director of News and Information at Robert Morris University in Pittsburgh, PA** for an interesting discussion about a passion he has for Skyscrapers. His book, "MultiStories" is a work of love and details how skyscrapers shaped modern America. Mark also talks about the tours of skyscrapers in Pittsburgh that he conducts and the interesting people that he meets on them.

Episode 48:

The Amazing Cities & Towns Podcast with Jim Hunt welcomes **John Brenner, the Executive Director designate of the Pennsylvania Municipal League** to discuss his role as the incoming director of the league and what he hopes to accomplish in this important role. John is the former Mayor of York, Pennsylvania, and is one of the up-and-coming league directors across the country. John's interesting stories about local government are heartwarming and bring a message to all in local government.

Episode 49:

The Amazing Cities & Towns Podcast with Jim Hunt and his guest, **Matt Zone, former Cleveland City Councilmember and past President of the National League of Cities.** Matt wrote the foreword for The Amazing City-7 Steps to Creating an Amazing City and has an incredible legacy of public service following both his father and mother as members of the Cleveland City Council. Matt's love for his city comes through in this interesting episode.

Episode 50:

The Amazing Cities & Towns Podcast with Jim Hunt discusses building a Brand for your City with **Jeni Bukolt, founder of Haven Creative**, a branding and communication expert in Charlotte, North Carolina.

Episode 51:

The Amazing Cities & Towns Podcast with Jim Hunt and his guest, **Melodee Colbert-Kean, former Mayor and current member of the Joplin, Missouri City Council and Past President of the National League of Cities.** Melodee is also a successful entrepreneur and community activist. Melodee talks about her journey to local government and how she came to lead the National League of Cities. Jim also discusses the devasting tornado that struck Joplin in 2011 and how the city recovered from the loss of 158 lives and over $2.5 Billion in damages. A great learning experience!

Episode 52:

The Amazing Cities & Towns Podcast with Jim Hunt discusses "Volunteering to Make Your City Amazing" with **Wayne Worth, a councilman in Clarksburg, West Virginia** who has blazed a trail with his incredible story of public service. Wayne was adopted as a young child in West Virginia and has spent his life working on behalf of his state. He will travel hundreds of miles to help poor families recover from a flood or spend his weekends on neighborhood cleanups that have made a difference in his community.

Episode 53:

The Amazing Cities & Towns Podcast with Jim Hunt features **former Rock Hill, South Carolina Mayor Doug Echols**. Doug was featured in Jim's book, "The Amazing City-7 Steps to Creating an Amazing City" and tells his story of a downtrodden city's rebirth as a sports and recreation center outside of Charlotte, North Carolina. Doug led the way to building a BMX bicycle facility to his town that now hosts international competitions as well as providing an amazing facility for local citizens.

Episode 54:

The Amazing Cities & Towns Podcast with Jim Hunt discusses with the podcast's producer, Matt James, a new partnership with Bearing Advisors to sponsor the podcast. Bearing Advisors works with local governments and clients to create innovative Public/Private Partnerships, and they are a perfect complement to the Amazing Cities and Towns Podcast. Looking for great things to come!

Episode 55:

The Amazing Cities & Towns Podcast with Jim Hunt welcomes **Kevin Samy, a senior advisor for the Obama Administration and now working with a company that is providing air quality equipment to businesses and local governments**. Kevin explains the U.S. Green Building Council's program WELL program and how to obtain this certification.

Episode 56:

The Amazing Cities & Towns Podcast with Jim Hunt welcomes **David Fivecoat, leadership consultant,** to the podcast to discuss his leadership work based on his twenty-four-year career with the United States Army. Colonel Fivecoat served as an infantry officer, leading men and women during operations in Kosovo and Bosnia, three combat tours in Iraq, and a combat tour in Afghanistan. He closed out his career leading the gender integration of the U.S. Army's Ranger School in Fort Benning, Georgia. Jim and David have an interesting discussion on leadership and innovation.

Episode 57:

The Amazing Cities & Towns Podcast with Jim Hunt begins a four-part series on Downtown Revitalization. Jim and Amazing Cities Podcast producer, Matt James discuss the various components of Downtown Revitalization. This insight is from Jim's lifelong work with cities and towns and the lessons learned.

Episode 58:

The Amazing Cities & Towns Podcast with Jim Hunt continues the series on Downtown Revitalization with **Matt James, producer of the Amazing Cities and Towns Podcast.** Matt talks to Jim about the existing resources available to cities and towns and how to go about finding the ideas and money to complete a successful downtown revitalization.

Episode 59:

The Amazing Cities & Towns Podcast with Jim Hunt continues with Part 3 of the Series on Downtown Revitalization with show producer,

Matt James. Matt and Jim talk about Jim's recent travel through over 30 cities and towns and the lessons he learned on both the successes and failures of today's downtown retail districts.

Episode 60:

The Amazing Cities & Towns Podcast with Jim Hunt concludes the 4-Part Series on Downtown Revitalization with a discussion about federal resources and what cities and towns can do to find money through some of the COVID-19 funds that are available. The secret to these funds is you must know where to look and who to contact. Matt James, producer interviews Jim Hunt.

Episode 61:

The Amazing Cities & Towns Podcast with Jim Hunt gives an update on his visits to several State Municipal League Conferences and the National League of Cities Conferences. As COVID ends, the nation slowly recovers, and state and national local government conferences are anxious to see people attending conferences.

Episode 62:

The Amazing Cities & Towns Podcast with Jim Hunt explores The Future of Electric Vehicles in Amazing Cites with **Rap Hankins, a former council member from Trotwood, Ohio, and now the president of Drive Electric Dayton.** Rap speaks about his advocacy work for the adoption of EVs in disadvantaged communities and the lack of charging stations throughout the country. Rap lives his passion by owning an EV and traveling throughout the country spreading the word.

Episode 63:

The Amazing Cities & Towns Podcast with Jim Hunt and the kick-off episode for 2022. Jim and his producer, Matt James discuss the highlights of the 2021 Podcast season and the exciting guests and topics lined up for 2022.

Episode 64:

The Amazing Cities & Towns Podcast with Jim Hunt and his guest, **Sarah Aquino, a Council member in Folsom, California** who made national news by helping bring attention to the labor shortage by taking a job in a restaurant in her city. She continues to assist a local eatery serving as a hostess several nights a week. A great story of public service and helping citizens and businesses survive a pandemic.

Episode 65:

The Amazing Cities & Towns Podcast with Jim Hunt welcomes **George Cuff to the podcast. George is a former mayor of Spruce Grove, Alberta, Canada, and is now a leading advisor, consultant, and author on governance and effective organization**. George gives practical advice and some interesting stories about his career in local government. He is a recognized expert and highly valued among his clients. A compelling episode!

Episode 66:

The Amazing Cities & Towns Podcast with Jim Hunt discusses Economic and Community Development **with Jim Byard, former Mayor of Prattville, Alabama, and head of the Alabama Department of Economic and Community Development under Governor Robert Bentley.** Jim puts a face on economic and community development and has a practical approach to helping local governments succeed.

Episode 67:

The Amazing Cities & Towns Podcast with Jim Hunt and his guest, **Ms. Ansley Fender, CEO of Atlas Solutions of Bloomington, Indiana**. Ansley tells her interesting and emotional story of being a violinist, having an injury that forced her to change fields and becoming an entrepreneur. She formed her new company as part of a business incubator in Bloomington and has been an enthusiastic advocate for cities and towns working to nurture young businessmen and women. A Great episode!

Episode 68:

The Amazing Cities & Towns Podcast with Jim Hunt features **Anne Uecker, a former city clerk and now a professional speaker** who uses her experience in local government to present entertaining and informative speeches and presentations throughout the United States and Canada. Jim met Anne at a Municipal Clerk's Conference in Wheeling, WV, and knew he had to have her on the podcast. A great personality and plenty of great information.

Episode 69:

The Amazing Cities & Towns Podcast with Jim Hunt welcomes **former Olympian, Amy Gamble** to talk about her upcoming book, "Unsilenced", a memoir of healing from trauma. Amy tells her compelling story of a fall into mental illness and the steps she continues to take to reclaim her life. One of our best episodes and hopefully helpful to our audience who may be struggling or have a friend or family member struggling with mental health.

Episode 70:

The Amazing Cities & Towns Podcast with Jim Hunt. This episode features **Aaron Hamlin, the director of the Center for Election Science** to discuss 'ranked' voting and how he feels this innovative way of voting is an improvement on the current system and allows for a better way to arrive at a fair and accurate result. The system is being used in Fargo, North Dakota and Aaron is looking to add other cities and towns to the list of communities that use ranked voting.

Episode 71:

The Amazing Cities & Towns Podcast with Jim Hunt discusses the NLC Congressional Cities Conference recently held in Washington, DC. The NLC Conference moved from its longtime venue to the Marriott Marquis which is adjacent to the Washington Convention Center and gave an entirely new feel to the conference. Jim recapped the important federal advocacy issues discussed at the conference and some of the new things facing cities and towns in the post-Covid era.

Episode 72:

The Amazing Cities & Towns Podcast with Jim Hunt discusses Improving Your City's Online Image. With Zoom and other online meeting platforms becoming a part of local government, what rules and safeguards should cities put in place to deal with this 'online world'. An interesting discussion about a new phenomenon in local government.

Episode 73:

The Amazing Cities & Towns Podcast with Jim Hunt welcomes **Roger Kemp, a veteran city manager, author, and thought leader on local government.** Roger talks about the value of having an educational institution in your community and the best ways to maximize the relationship. A great discussion on Town-Gown relations.

Episode 74:

The Amazing Cities & Towns Podcast with Jim Hunt welcomes back **John Brenner, the Executive Director of the Pennsylvania Municipal League and former Mayor of York, Pennsylvania.** This is an interesting episode on a variety of topics that talk about funding infrastructure, finding resources, and legislative lobbying. John is a rock star and always is entertaining while sharing valuable information.

Episode 75:

The Amazing Cities & Towns Podcast with Jim Hunt welcomes **Dr. Barry E. Truchill, author of "The Politics of Local Government"** Dr. Truchill gives some excellent analysis of issues like historic preservation and local zoning that is fascinating. He uses the Joni Mitchell song lyric, 'I've looked at life from both sides now' to describe his work in local government.

Episode 76:

The Amazing Cities & Towns Podcast with Jim Hunt speaks with **Ann Macfarlane, a registered parliamentarian and owner of The Jurassic Parliament, a consulting company** that focuses on teaching parliamentary procedure to local government leaders and others. Her fascinating career as a Foreign Service Officer and her work in Pakistan and then, Bangladesh was a bonus to the discussion.

Episode 77:

The Amazing Cities & Towns Podcast with Jim Hunt welcomes **Paul Helmke, former Mayor of Fort Wayne, Indiana, and Past President of the U.S. Conference of Mayors**. Paul is now a professor at Indiana University and is sharing his wisdom with students as a way of giving back to a profession he loves. Paul also led the Brady Center to prevent gun violence for five years and shares how a Republican mayor was tasked with addressing the topic of gun violence after the tragic shooting of President Ronald Reagan and the life-altering injuries to Reagan's Press Secretary Jim Brady.

Episode 78:

The Amazing Cities & Towns Podcast with Jim Hunt welcomes **Tom Fountaine, City Manager of State College, Pennsylvania,** and the home of Penn State University. Tom reflects on his long career in State College and the challenges of having thousands of college students living in your community. We also discussed the unique situation created when Penn State plays a big game at home and the city is tasked with controlling over 100,000 fans who enter the city for the game.

Episode 79:

The Amazing Cities & Towns Podcast with Jim Hunt welcomes **Travis Blosser, the Executive Director of the West Virginia Municipal League** to discuss his journey to become the leader of West Virginia's cities and towns and the biggest issues facing them, being in a rural state. Travis has the distinction of being an elected city council member, a city manager in two West Virginia cities, and then joining the Municipal League and finally taking over the head job. An interesting journey!

Episode 80:

The Amazing Cities & Towns Podcast with Jim Hunt welcomes **Leon Andrews, Jr., CEO and President of Equal Measure and former leader of the National League of Cities' Race, Equity, and Leadership initiative**. The discussion centered on the continuing struggle for equality and racial justice in America and on the heels of the George Floyd murder, what the future looks like for his work.

Episode 81:

The Amazing Cities & Towns Podcast with Jim Hunt gives a wide-ranging discussion about the progress of the podcast and highlights past episodes and the topics they covered.

Episode 82:

The Amazing Cities & Towns Podcast with Jim Hunt welcomes **Nicole Rongo, Vice-President of Government Relations and Strategic Partnerships at CGI Digital, an Enterprise Partner of the National League of Cities**. Nicole gave a history of how CGI Digital grew from a small community map sales company in Rochester, New York to become a multi-media digital leader in innovative public/private partnerships with cities and towns across America.

Episode 83:

The Amazing Cities & Towns Podcast with **Jim Hunt welcomes Jeff Towery, President of the International City/County Management Association and City Manager of McMinnville, Oregon.** Jeff talks about his plans for his presidential year with ICMA. He gives an insight into ICMA's work on both a domestic and international stage and the growth of the organization.

Episode 84:

The Amazing Cities & Towns Podcast with Jim Hunt welcomes **Manny Teodoro, Co-Author of The Profits of Distrust, an excellent book about water, public confidence in government,** and where we go from here. We discuss the ongoing water infrastructure crisis in the country with examples like Flint, Michigan, Jackson, Mississippi, and the Camp Lejeune contaminated water problem. An interesting conversation from an expert in the field who talks openly and bluntly about solving these problems.

Episode 85:

The Amazing Cities & Towns Podcast with Jim Hunt welcomes **Greg Cochran, Executive Director of the Alabama League of Municipalities** who discusses the growth of the Alabama league and how they have grown in staff and expertise to meet the emerging issues in municipal government. Greg's long career with the league gives him a rare perspective on how cities and towns need to grow in leadership development to meet the expectations of the public.

Episode 86:

The Amazing Cities & Towns Podcast with Jim Hunt welcomes **Adam McGough, Council member for District 10 in the city of Dallas, Texas.** Adam gives a unique perspective on the duties and responsibilities of a large city council member. Adam recounts his realization that he was regularly dealing with issues that ran into the billions of dollars and how it changes your mindset, thinking in those huge numbers. A rare look at the inside of big city politics!

Episode 87:

The Amazing Cities & Towns Podcast with Jim Hunt welcomes **Dr. Kimberly Nelson of the School of Government at the University of North Carolina** for a discussion on the evolution of forms of local governments and the why's behind some things that we take for granted. Dr. Nelson is a researcher, and her unique insights will give our listeners some thoughts and ideas they may not have looked at in that way.

Episode 88:

The Amazing Cities & Towns Podcast with Jim Hunt welcomes **Mayor Steve Patterson of Athens, Ohio, and a member of the National League of Cities Board of Directors.** Steve recounts leaving his full-time professor position to serve as Mayor and trading lifetime tenure for the fickleness of the voters, every four years. Steve is an interesting person and a fantastic storyteller. This entertaining episode is sure to be a favorite.

Episode 89:

The Amazing Cities & Towns Podcast with Jim Hunt welcomes **former Augusta, Georgia City Manager Janice Jackson.** Janice is also a podcast host and has a great personality as both a host and a guest. Janice talks about her experiences in Augusta and we even talked about the famous golf course that hosts The Masters golf tournament yearly. Janice also talks about her experiences seeing hundreds of homeless persons in Portland, Oregon when she attended an ICMA conference and the growing problem throughout the country.

Episode 90:

The Amazing Cities & Towns Podcast with Jim Hunt welcomes **Allentown, Pennsylvania Mayor Matt Tuerk** for a discussion on his community and his unique perspective of his job leading one of Pennsylvania's largest cities. Matt points out the fact that Allentown is growing and changing demographically and is now a majority Latino city as of the 2020 census. Mayor Tuerk also relates the story of a "Punk Rock" kid who became the mayor, with tattoos and all. He talks about finding mentors among nearby mayors and picking up some good ideas at the Pennsylvania Municipal League meetings.

Episode 91:

The Amazing Cities & Towns Podcast with Jim Hunt welcomes **Joe Delong, Executive Director of the Connecticut Conference of Municipalities.** Joe is a former football star from West Virginia University and a former West Virginia State legislator who took a career switch and headed north to lead an organization helping cities and towns. Joe is a great storyteller and his journey to public service has deep roots and he tells it compellingly. Don't miss it!

Episode 92:

The Amazing Cities & Towns Podcast with Jim Hunt welcomes **New York Times Best Selling Author, Sam Quinones** to the podcast. He discusses his latest book, "The Least of Us" where he investigates the unbelievable impact that fentanyl and meth is having in communities across the United States. His first book, "Dreamland" explored the worldwide opioid epidemic, and he is considered one of the leading authorities on the scourge of drugs impacting American cities and towns. A must listen to episode for anyone interested in the future of our cities and towns.

Episode 93:

The Amazing Cities & Towns Podcast with Jim Hunt welcomes **Mariel Reed, the Co-Founder and CEO of Pavillion**, an organization working to streamline procurement for local governments. This interesting discussion centers around the challenges of procurement processes for local government and making it easier to share the work that they've already done. Mariel tells the story of her husband, making her promise not to bring up the "P" word when they go out with others, but trust me, she is great at making a mundane subject interesting. The "P" word is procurement!

Episode 94:

The Amazing Cities & Towns Podcast with Jim Hunt welcomes two young local government leaders who are working on a program called Local Government 2030. **Mike Zeller from Shrewsbury, Massachusetts, and Bianca Alvarez from San Antonio, Texas** talk about the National Academy of Public Administration's 12 Grand NAPA Challenges. The energy of these youthful leaders is infectious, and they describe their work and what they believe will be the results. If you want a peek at the future of local government, these two young people will inspire you.

Episode 95:

The Amazing Cities & Towns Podcast with Jim Hunt welcomes **Henry Bouchot to discuss his book, "A Millennial's Guide to Running for Office-How to Get Elected without Kissing the Ring!"** This intriguing title is the beginning of an interesting conversation on local politics for the millennial generation. Funny and informational!

Episode 96:

The Amazing Cities & Towns Podcast with Jim Hunt discusses the "Great Resignation" with **Joe Mull, author of Employaty**, a book about what leaders must do every day to create teams that work hard, get along, and produce results. Joe is a member of the National Speakers Association and travels throughout the United States as a professional speaker and consultant. A great episode for those wondering how local governments can recruit and retain excellent employees.

Episode 97:

The Amazing Cities & Towns Podcast with Jim Hunt welcomes **Selma, Alabama Mayor James Perkins Jr.** to discuss the opportunities and challenges of being the first African American mayor in one of the Civil Rights movement's most pivotal moments. Mayor Perkins is a thoughtful leader who came back home from a successful technology career to run for mayor in the shadow of the Edmund Pettis Bridge, the site of the Bloody Sunday confrontation that left many injured and showed the nation the brutality of the fight for freedom for African Americans. One of the best episodes of the podcast!

Episode 98:

The Amazing Cities & Towns Podcast with Jim Hunt welcomes **Dr. Terry Christiansen a professor at San Jose State University**. Professor Christiansen brings one of the most well-rounded discussions about local government and his outstanding work teaching thousands of students valuable life lessons on the value of getting involved in local government. He discusses the demise of the local press and the limitations of governing in a Zoom world.

Episode 99:

The Amazing Cities & Towns Podcast with Jim Hunt and his guest **Julia Payson discusses her book, "When Cities Lobby: How Local Governments Compete for Power in State Politics"**. Julia has done extensive studies on how state governments impact cities and towns and why cities must lobby to compete in a polarized environment.

Episode 100:

Amazing Cities & Towns Podcast with Jim Hunt and his guest speaks with **2023 NLC President and Mayor of Tacoma, WA Victoria Woodards** about her successful year as president and how it affected her city of Tacoma. Victoria talks about her interesting path to becoming mayor and how she has worked with the private sector to build exciting Public-Private Partnerships that have shown great success in her city.

Episode 101:

The Amazing Cities & Towns Podcast with Jim Hunt talks about reaching 100 episodes of the podcast.

Episode 102:

The Amazing Cities & Towns Podcast with Jim Hunt and his guest, **Ken Duran, a Senior Economic Development Advisor for HdL Companies** discuss the work that HdL is doing to assist cities in using data for economic development. He is a former city manager, assistant city manager, and parks director and understands how cities can use data.

Episode 103:

The Amazing Cities & Towns Podcast with Jim Hunt and **Carolyn Coleman, Executive Director of the League of California Cities and former Chief Lobbyist for the National League of Cities** discuss the opportunities and challenges of advocating for cities and towns. We also discuss issues such as homelessness and drug addiction, and what California cities and towns are doing to address them.

Episode 104:

The Amazing Cities & Towns Podcast with Jim Hunt and **Patrick Leddin, author of the "5 Week Leadership Challenge"** discusses his interesting life story and his Wall Street Journal best-selling book, "The 5 Week Leadership Challenge". Patrick's organization works with a wide variety of clients and achieves success through its series of 35 Action Steps. Valuable Leadership Advice!

Episode 105:

The Amazing Cities & Towns Podcast with Jim Hunt and his guest, **Lee Feldman, former City Manager of Fort Lauderdale, Florida, and a Past President of the International City/County Management Association**. Lee now works with Zencity and we discuss how Zencity can help state and local governments obtain community input, understand it, and use it to make effective decisions.

Episode 106:

The Amazing Cities & Towns Podcast with Jim Hunt and **Dr. Maria Church discusses her book, "Love-Based Leadership"**. Dr. Church leads an organization that works with many cities and towns to build leadership skills and improve communication and interpersonal skills.

Episode 107:

The Amazing Cities & Towns Podcast with Jim Hunt and **Sam Lusk, the Economic Development Director for the City of Princeton, West Virginia**. Sam is a young professional with experience working for a United States Congressman and he speaks about his exciting work in a small Appalachian community and the successes he is having.

Episode 108:

The Amazing Cities & Towns Podcast with Jim Hunt and **Scott Hutcheson, the executive director of E Pluribus Unum** have a candid conversation about racial equality in the South. A fascinating discussion!

Episode 109:

The Amazing Cities & Towns Podcast with Jim Hunt and **2024 NLC President David Sander** discuss the upcoming 100th Birthday of the National League of Cities and the gala celebration planned for the City Summit in Tampa, Florida.

Episode 110:

The Amazing Cities & Towns Podcast with Jim Hunt and **Shawana Cross, a council member in Westover, West Virginia, and an urban bee advocate.** Shawana got involved in local government when a code inspector told her she could not have bee hives in the city and she went to work educating the city council and getting an ordinance passed to permit them. She then got herself elected to the city council. A fun and informative discussion!

Episode 111:

The Amazing Cities & Towns Podcast with Jim Hunt and his guest **Barton Drake, Principal of HED, a design, architecture, and engineering company.** HED has done some great work in Garland, Texas turning a city hall rehab project into a beautiful Main Street revitalization. Barton understands local government and shares his insights on building Amazing cities.

Episode 112:

The Amazing Cities & Towns Podcast with Jim Hunt and his guests, **Colleen Hilton, CEO of Alli Connect and Angelo Consilo, President of FOP Lodge 89 in Prince George's County, MD** discuss the issue of mental health resources for first responders and the innovative way that Colleen's company, Alli Connect can match trained professionals with first responders in need. An important show that all local government officials should listen to.

Episode 113:

The Amazing Cities & Towns Podcast with Jim Hunt and **Chris May of Advantage Technologies** discusses a critical topic for local government officials, cybersecurity. Chris explains the challenges facing local government and some basic security steps that every city should take.

Episode 114:

The Amazing Cities & Towns Podcast with Jim Hunt, and his guest, **Dr. Seth Kaplan author of "Fragile Neighborhoods"** discusses the vision behind his book. Seth explains how to repair American Society, one neighborhood at a time. Great show!

Episode 115:

The Amazing Cities & Towns Podcast with Jim Hunt, and his guest **Glenn Akramoff, Author of The Human Centered Team** talk about leading teams with a shared vision and purpose for optimal performance.

Episode 116:

The Amazing Cities & Towns Podcast with Jim Hunt, and his guest, **Mackenzie Bent, the manager of Global Development with CitiIQ.** An interesting discussion about global and regional development challenges.

Episode 117:

The Amazing Cities & Towns Podcast with Jim Hunt, and his guest, **Woody Woodworth, CEO of The Brick Painters**. Jim and Woody have a fun conversation about cities and towns using an innovative and inexpensive process to restore the brick buildings.

Episode 118:

The Amazing Cities & Towns Podcast with Jim Hunt, and his guest, **Rob Parker, the President of the Town of Trilith**. Rod and Jim discuss "New Urbanism", and his entrepreneurial journey in the film industry, and why he decided to build a town.

Episode 119:

The Amazing Cities & Towns Podcast with Jim Hunt, and his guest, **Anna Lowder who is the founder of the Town of Hampstead in Montgomery, Alabama**. Anna is active in the "New Urbanism" movement, and she discusses the innovative housing development trends that she brings from her time living in London.

Episode 120:

The Amazing Cities & Towns Podcast with Jim Hunt, with **Shae Strait, the City Planner for the city of Fairmont, West Virginia**. Shae discusses brownfield redevelopment and an insightful discussion of the "Strong Towns" movement.

Episode 121:

The Amazing Cities & Towns Podcast with Jim Hunt, with **Jason Burnett, the CEO of Crosswalk Labs**, a company that tracks greenhouse gas emissions across the United States. The former mayor of Carmel-by-the-Sea has an extensive record of working for a cleaner environment.

Episode 122:

The Amazing Cities & Towns Podcast with Jim Hunt, and **Steve Patterson, Athens, Ohio Mayor and 2nd Vice-President of the National League of Cities** discuss a recent trip to Ukraine that he made.

An in-depth conversation about the realities of local government on a global basis.

Episode 123:

The Amazing Cities & Towns Podcast with Jim Hunt, and **Diane Kalen-Sukra, best-selling author of "Save Your City: How Toxic Culture Kills Community and What to do About It",** discusses the increasing toxicity present in local government.

Episode 124:

The Amazing Cities & Towns Podcast with Jim Hunt and his guest, **Melissa Hinojosa, Councilmember in Madisonville, Texas** discusses the challenges and opportunities for young people entering local government.

Episode 125:

The Amazing Cities & Towns Podcast with Jim Hunt and his guest, **Erik Caldwell, co-founder and President/CEO of Metropolis IQ** and seasoned AI data product leader with a strong background in local government.

Episode 126:

The Amazing Cities & Towns Podcast with Jim Hunt and his guest, **Adrian Brown** from the Centre for Public Impact discuss addressing complex issues with community members.

Episode 127:

The Amazing Cities & Towns Podcast with Jim Hunt and his guest, Andy Nickerson, CEO of HdL Companies of Brea, CA. Andy discusses HdL's recent expansion and the growth of the data analytics that is fueling new innovations in providing data to cities and towns. Also discussed the upcoming ICMA Conference to be held in Pittsburgh, PA.

About the Author

Jim Hunt is a former Mayor and City Councilmember with over 27 years of experience in
local government, including serving as President of the National League of Cities. Recognized as a leading voice in municipal leadership,
Jim has dedicated his career to helping cities
thrive through innovation and entrepreneurial
thinking.

He is the author of the highly successful
book *The Amazing City: 7 Steps to Creating an
Amazing City* and host of the *Amazing Cities and
Towns Podcast*, where he shares insights on governance and city leadership. As a consultant and speaker, Jim works with local governments to
address complex issues such as leadership coaching, strategic planning,
council dysfunction, and citizen engagement. He also engages with the
business community to develop pathways into local government and
provide insight on successful public/private partnerships.

In 2006, Jim was selected as *Municipal Leader of the Year* by Amer-
ican City & County Magazine for his work creating the *Partnership for
Working Towards Inclusive Communities*. In 2011, Jim Hunt led a National
League of Cities (NLC) delegation of U.S. representatives to Durban,
South Africa to sign a Memorandum of Understanding with the South
Africa Local Government Association (SALGA) strengthening and
promoting local governance.

Jim resides in West Virginia with his wife, where he continues to
consult, write, and speak on local government issues.

ARE YOU PLANNING A MEETING, CONFERENCE, OR EVENT?

Invite Jim Hunt to be your Keynote Speaker

"Jim Hunt is one of America's leading thinkers on improving America's cities. His passion and energy are evident from the moment he steps on stage. You won't be disappointed!"

Lisa Dooley, Former Executive Director
West Virginia Municipal League.

For speaking engagements, consulting, or media inquiries, Jim Hunt brings unmatched expertise and insight to your organization or event. To book Jim or to inquire about bulk book purchases for conferences, training programs, or giveaways, please visit www.amazingcities.org or contact him at *jimhunt@amazingcities.org*. Take advantage of his unique perspective to inspire innovation and entrepreneurial leadership in your city or organization.